KEY TOPICS IN BRAIN RESEARCH

Edited by A. Carlsson, P. Riederer,
H. Beckmann, T. Nagatsu,
and S. Gershon

T. Nagatsu, H. Narabayashi, and M. Yoshida (eds.)

Parkinson's Disease. From Clinical Aspects to Molecular Basis

Springer-Verlag Wien New York

Prof. Dr. Toshiharu Nagatsu
Department of Neurochemistry, Fujita Health University,
Aichi, Japan

Prof. Dr. Hirotaro Narabayashi
Neurological Clinic, Tokyo, Japan

Prof. Dr. Mitsuo Yoshida
Department of Neurology, Jichi Medical School, Tochigi-ken, Japan

Printed on acid-free paper
Product Liability: The publisher can give no guarantee for information about drug dosage and application thereof contained in this book. In every individual case the respective user must check its accuracy by consulting other pharmaceutical literature. The use of registered names, trademarks, etc. in this publication does not imply, even in the absence of a specific statement, that such names are exempt from the relevant protective laws and regulations and therefore free for general use.

With 52 Figures

ISSN 0934–1420

ISBN-13: 978-3-211-82272-2   e-ISBN-13: 978-3-7091-9146-0
DOI: 10.1007/978-3-7091-9146-0

Preface

This book reviews the recent advances in the research on Parkinson's disease. It contains review articles from basic to clinical researches including the historical introduction and the molecular biological approach to Parkinson's disease.

Parkinson's disease is the most representative, age-related neurodegenerative disease. It is clinically characterized by movement disorders such as muscle rigidity, akinesia and tremor. The elucidation of its biochemical and molecular mechanisms has rapidly been progressing and is expected to contribute to the understanding of normal brain aging in general.

This project is supported by a Grant-in-Aid for Scientific Research on Priority Areas, "Molecular Biology of the Motor System", Ministry of Education, Science and Culture, Japan, for 3 years from 1987 to 1989. We are grateful for the support. These reviews are a part of the works which have been supported by this Grant-in-Aid.

The characteristics of this research project on Parkinson's disease are interdisciplinary approach from basic, molecular biology to clinical medicine, and the molecular biological approach is expected to be the most promising for the elucidation of the pathogenesis. The collaboration and discussion between basic and clinical researchers in this Priority Area Project has been efficient and productive.

We hope that this book can mark another new milestone in the studies on Parkinson's disease and in neuroscience.

We are very grateful to Nippon Schering for their generous support for the publication of this book.

Last but not least, we thank Springer-Verlag Wien New York for the excellent production of this book and the excellent cooperation.

June 1990   T. NAGATSU, H. NARABAYASHI, and M. YOSHIDA

Contents

Contents

Tyrosine hydroxylase in relation to Parkinson's disease: a historical overview and future prospect

T. Nagatsu

Department of Biochemistry, Nagoya University School of Medicine, Nagoya, Japan

Summary

Reduction in tyrosine hydroxylase (TH), the regulatory enzyme for dopamine (DA) biosynthesis, in the nigrostriatal DA region is a characteristic change in Parkinson's disease. TH was discovered in 1964 as the first and regulatory enzyme in DA biosynthesis, and the specific decrease of TH activity in the nigrostriatal DA region of parkinsonian brains was established around 1975. It has been generally assumed that the decrease in TH activity in parkinsonian brain is due to a reduction in TH protein as a result of cell death of the nigrostriatal DA neurons by unidentified mechanisms. In fact, we found TH activity and TH protein to be decreased in parallel in parkinsonian brain. However, since the discovery of multiple forms of human TH in 1987, some molecular changes in TH protein in Parkinson's disease are speculated. This speculation may be supported by the increase in homospecific activity (enzyme activity/enzyme protein) of residual TH in parkinsonian brain, in contrast to constant homospecific activity of TH in the striatum of parkinsonian mice treated with 1-methyl-4-phenyl-1,2,3,6-tetrahydropyridine (MPTP). A molecular biology approach to TH in Parkinson's disease is expected to reveal the molecular pathogenesis.

Introduction

Parkinson's disease is a neurodegerative motor disease related to brain aging, and is clinically characterized by muscle rigidity, akinesia, and tremor. The characteristic pathological and biochemical changes in Parkinson's disease are a severe cell loss of dopamine (DA) cell bodies in the substantia nigra and a severe decrease of DA in the striatum (Bernheimer et al., 1973).

DA is a catecholamine neurotransmitter with noradrenaline and adrenaline, and is synthesized from tyrosine by the following pathway: tyrosine → 3,4-dihydroxyphenylalanine (DOPA) → DA (Fig. 1). Tyrosine hydroxylase (TH) (1) and aromatic L-amino acid decarboxylase (DOPA decarboxylase) (2) catalyze the two biosynthetic steps. TH was discovered in 1964 as a pteridine-requiring monooxygenase (Nagatsu et al., 1964), and was determined to be the rate-limiting step in catecholamine biosynthesis in $vivo$ (Levitt et al., 1965).

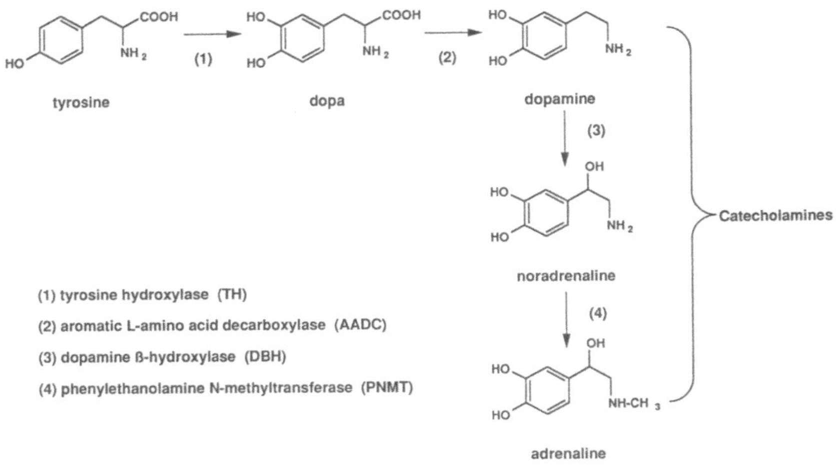

Fig. 1. The pathway and the related enzymes of catecholamine biosynthesis

A significant reduction in TH activity in the nigrostriatal DA region of the postmortem brains from the patients with Parkinson's disease was found around 1975 (Lloyd et al., 1975; Nagatsu, 1975; McGeer and McGeer, 1976; Nagatsu et al., 1977). In the case of biochemical studies on postmortem human brains, only the changes in TH at the time of death are examined. The changes in TH prior to, at the onset of, or during the progress of Parkinson's disease are unknown. The decrease in TH activity in parkinsonian autopsy brains is assumed to be a result of the cell loss at the late stage of the disease.

1-Methyl-4-phenyl-1,2,3,6-tetrahydropyridine (MPTP) produces parkinsonism in humans (Davis et al., 1979; Langston et al., 1983), monkeys (Burns et al., 1983), and mice (Heikkila et al., 1984). MPTP-induced parkinsonism appears to be clinically, pathologically and biochemically similar to Parkinson's disease.

We used MPTP-treated mice to follow the changes in TH at the early and late stages and have compared the results with the changes in TH in human parkinsonian brains.

Human tyrosine hydroxylase (TH): the multiple forms

TH catalyzes the first step of catecholamine biosynthesis, the hydroxylation of L-tyrosine to L-DOPA (Fig. 1). TH requires a tetrahydropteridine cofactor (Nagatsu et al., 1964). The natural pteridine cofactor is (6R)-L-*erythro*-tetrahydrobiopterin (Brenneman and Kaufman, 1964; Matsuura et al., 1985). As shown in Fig. 2, TH hydroxylates L-tyrosine to L-DOPA with molecular oxygen and tetrahydrobiopterin to produce quinonoid dihydrobiopterin and

H_2O. In tissues, quinonoid dihydrobiopterin is reduced back to tetrahydro-biopterin by NADH-requiring dihydropteridine reductase and NADH (Togari et al., 1983) (Fig. 2). Fe^{2+} activates TH activity in various animals (Nagatsu et al., 1964). Human TH is the most dependent on Fe^{2+} (Nagatsu et al., 1972; Ishii et al., 1990). Human TH was purified from adrenals (Kojima et al., 1984; Mogi et al., 1984) and brain (Mogi et al., 1986), and the enzyme purified from human adrenals was found to be markedly activated by exogenous Fe^{2+} (Ishii et al., 1990).

Both cDNA (Grima et al., 1987; Kaneda et al., 1987; Kobayashi et al., 1987) and genomic DNA (O'Malley et al., 1987; Kobayashi et al., 1988) of human TH has been cloned and the nucleotide sequence has been determined (Fig. 3, 4). The results of Southern blot analysis and the nucleotide sequence of the human TH genomic DNA indicate that four similar but distinct mRNAs encoding human TH are produced through alternative splicing from a single gene. These mRNAs are constant for the major part but are distinguishable from one another as to the insertion/deletion of 12- and 81-bp sequences near the N-terminus. The type 2 and 3 mRNAs contain the 12- and 81-bp insertion sequences of type 4, respectively, into the sequence of type 1. In type 4 mRNA, a 93-bp sequences composed of the 12- and 81-bp sequences is inserted at the same position of type 1. Since this insertion does not alter the reading frame of the protein-coding region, type 4 cDNA codes the longest TH molecule.

Type 1 human TH mRNA contains the coding region of 1491-bp encoding 497 amino acids, which has an estimated molecular weight of 55533. Numbers of amino acids (and molecular weight) of other types of human TH are 501 (Mr = 55973) for type 2,524 (Mr = 58080) for type 3, and 528 (Mr = 58521) for type 4, respectively.

The four types of human TH expressed in COS cells had similar Km values toward tyrosine and (6RS)-methyltetrahydropterin, but different homospecific

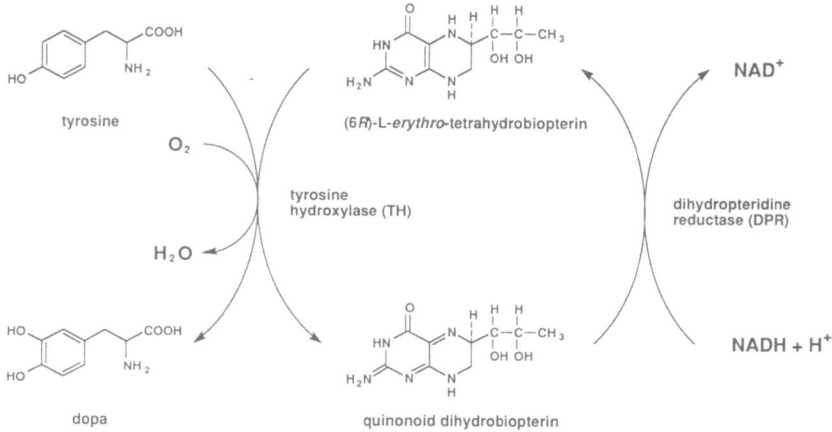

Fig. 2. The reaction of tyrosine hydroxylase (TH)

```
                                                                                    -1
                                                                              ACTGAGCC
                                                                                   120
ATGCCCACCCCGACGCCACCACGCCACAGGCCAAGGCGTTCCGCAGGCCGGTGTCTGAGCTGGACGCCAGACGAGGCCATCATGGTAAGAGGCGAGGGCGCCCGGGGCCCAGC
 M  P  T  P  D  A  T  T  P  Q  A  K  G  F  R  R  A  V  S  E  L  D  A  K  Q  A  E  A  I  M  V  R  G  Q  A  P  G  P  S
                                                                                   240
CTCACAGGCTCTCCGTGGCCTGGAACTGCAGCCCCAGCTGCATCCTACACCCCAAGGTCCCGCGTTCATTGGGCGCAGAGCCTCATCGAGGACGCCCAAGGAGCGG
 L  T  G  S  P  W  P  G  T  A  A  P  A  A  S  Y  T  P  T  P  R  S  P  R  F  I  G  R  R  Q  S  L  I  E  D  A  R  K  E  R
                                                                                   360
GAGGCGGCGGTGGCAGCAGCGGCCGCTGCAGTCCCTCGGAGCCCCTGGAGCCCTGTGGCCTTTGAGGAGGAGGAGAAGGGCGTGCTAAACCTGCTCTTCTCCCGAGG
 E  A  A  V  A  A  A  A  A  A  V  P  S  E  E  P  G  D  P  L  E  A  V  A  F  E  E  E  K  E  G  K  A  V  L  N  L  L  F  S  P  R
                                                                                   480
GCCACCAAGCCCTCGGCGCTGTCCCGTGCCGTGAAGGTGTTTGAGACGTTTGAGGCCAAAATCCACCATCTAGAGACCCGGCCCCAGCGGCCGCGGCGAGCTGGGGCCCCAACCTGGAG
 A  T  K  P  S  A  L  S  R  A  V  K  V  F  E  T  F  E  A  K  I  H  H  L  E  T  R  P  A  Q  R  P  R  R  A  G  Q  P  H  L  E
                                                                                   600
TACTTCGTGCGCCTCGAGGTGCGCCGAGGGGACCTGGCCGCGCTGCTCAGTGGTGTGCGCCAGGTTCAGAGTGAGGACGTGCGCAGCCCCGCAGGTCCCAAGGTCCCTGGTTCCCAAGAAA
 Y  F  V  R  L  E  V  R  R  G  D  L  A  A  L  L  S  G  V  R  Q  V  S  E  D  V  R  S  P  A  G  P  K  V  P  W  F  F  P  R  K
                                                                                   720
GTGTCAGAGCTGGACAAGTGTCATCACCTGGTCACCAAGTTCGACCCTGACCTGGACCACCCGGGCTTCTCGGACCAGGTGTACCGCCAGCGCAGGAAGCTGATTGCTGAGATC
 V  S  E  L  D  K  C  H  H  L  V  T  K  F  D  P  D  L  D  H  P  G  F  S  D  Q  V  Y  R  Q  R  R  K  L  I  A  E  I
                                                                                   840
GCCTTCCAGTACAGGCACGGCGACCCGATTCCCCGTGTGGAGTACACCGCCGAGGAGATTGCCACCTGGAAGGAGGTCTACACCACGCTGAAGGGCCTCTACGCCACGCACTGCGGG
 A  F  Q  Y  R  H  G  D  P  I  P  R  V  E  Y  T  A  E  E  I  A  T  W  K  E  V  Y  T  T  L  K  G  L  Y  A  T  H  A  C  G
                                                                                   960
GAGCACCTGGAGGCCTTTGCTTTGCTGGAGCGCTTCAGCGGCTACCGGGAAGACAACATATCCCCAGCTGAGGACGTCTCCCGCTTCCTGAAGGAGCGCACGGGCTTCCAGCTGCGGCCT
 E  H  L  E  A  F  A  L  L  E  R  F  S  G  Y  R  E  D  N  I  P  Q  L  E  D  V  S  R  F  L  K  E  R  T  G  F  Q  L  R  P
                                                                                  1080
GTGGCCGGCCTGCTGTCCGCCCGGGACTTCCTGGCCAGCCTGGCCTTCCGGGTGTTCCAGTGCACCCAGTATATCCGCCACGCGTCCTCGCCCATGCACTCCCCTGAGCCGGACTGCTGC
 V  A  G  L  L  S  A  R  D  F  L  A  S  L  A  F  R  V  F  Q  C  T  Q  Y  I  R  H  A  S  S  P  M  H  S  P  E  P  D  C  C
                                                                                  1200
CACGAGCTGCTGGGGCACGTGCCCATGCTGGCCGACCGCACCTTCGCGCAGTTCTCGCAGGACATTGGCCTGGCGTCCCTGGGGGCCTCGGATGAGGAAATTGAGAAGCTGTCCACGCTG
 H  E  L  L  G  H  V  P  M  L  A  D  R  T  F  A  Q  F  S  Q  D  I  G  L  A  S  L  G  A  S  D  E  E  I  E  K  L  S  T  L
                                                                                  1320
TCATGGTTCACGGTGGAGTTCGGGCTGTGTAAGCAGAACGGGGAGGTGAAGGCCTACGTCGGCGTGCTCTCCAGCTACGGGGAGCTGCTGCACTGCCTGTCTGAGGAGCCTGAGATT
 S  W  F  T  V  E  F  G  L  C  K  Q  N  G  E  V  K  A  Y  V  G  V  L  S  S  Y  G  E  L  L  H  C  L  S  E  E  P  E  I
                                                                                  1440
CGGGCCTTCGACCCTGCGGCCGCCGTGCAGCCGTACCAGGTACCAGCGTACTTCGTGTCTGAGAGCTTCAGTGACGCCAAGACAAGCTCAGGAGCTATGCCTCA
 R  A  F  D  P  E  A  A  A  V  Q  P  Y  Q  D  T  Y  Q  S  V  Y  F  V  S  E  S  F  S  D  A  K  D  K  L  R  S  Y  A  S
                                                                                  1560
CGCATCCAGCGGCCCCTTCTCCGTGAAGTTCGACCCGTACACGCTGGCCATCGACGTGCTGGACAGCCCCCAGGCCGTGCGGCGCTCCCTGGAGGGGTGTCCAGGATGAGCTGGACACCCTT
 R  I  Q  R  P  F  S  V  K  F  D  P  Y  T  L  A  I  D  V  L  D  S  P  Q  A  V  R  R  S  L  E  G  V  Q  D  E  L  D  T  L
                                                                                  1680
GCCCATGCTGAGTGCCTGTAGGTGCACGGCGTCCTGAGGGCCCTTCCCAACCTCCCCTGGTCCTGCACTGTCCCGACTGGGCTCAGGCCCTGGTCAGGGGCTGGGTCCTGCTCCGGTGCC
 A  H  A  L  S  A  I  G  ***
                                                                                  1800
CCCCATGCCCTCCCGCTGCCAGGCTCCCACGTCCCCTGCTTCTCAGCGCAACAGCTGTGTGTGCCCGTGGTGAGGTTGTGCTGGTGGTAGTGCTGGTCCTGCTCTGCTCCCAG
                                                                                  1860
GGTCCTGGGGCGTGCTGCACTGACCGCTCCGCCCTCCGTTCCCTGACACTGTGTGCCCAATCACCGTCACAATAAAAGAAACTGTGGTCTCT(A)n
```

Fig. 3. Nucleotide sequence and deduced amino acid sequence of type 4 human TH cDNA. Type 2 and type 3 human TH cDNA contain only the 12-bp sequence on the dotted line and the 81-bp sequence on the solid line, respectively. Type 1 human TH lacks the 93-bp sequence on the dotted and solid lines (Kaneda et al., 1987)

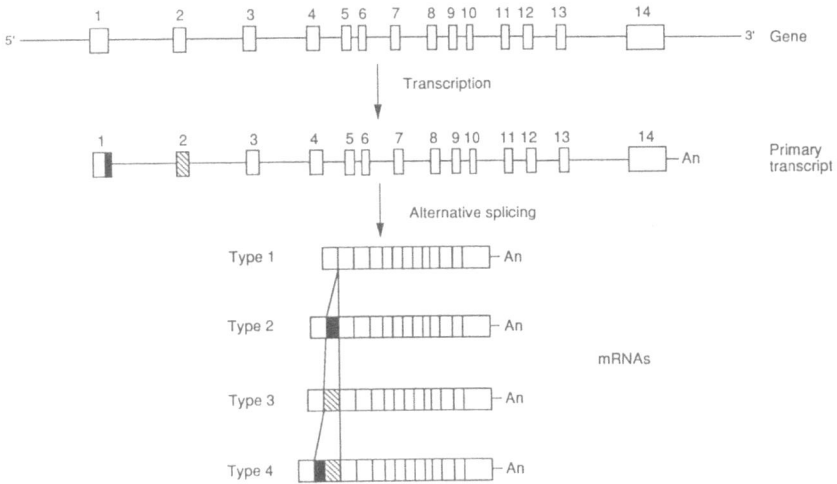

Fig. 4. Schematic illustration of regulation of the alternative splicing pathway producing the four types of human TH mRNA from the same primary transcripts. The 3′-terminal portion of exon 1, which corresponds to the 12-bp insertion sequence, is indicated by a filled box. The hatched box shows exon 2 that encodes the 81-bp insertion sequence (Kobayashi et al., 1988a)

activity (TH activity/TH protein), which was proposed as an immno-chemical index of enzyme by Rush et al. (1974). The type 1 enzyme had the highest homospecific activity, the values for the other enzymes ranging from about 0.3 to 0.4. Therefore, the insertion sequences appear to inhibit the TH activity (Kobayashi et al., 1988b).

We have recently found that the four types of human TH mRNA were expressed in the human brain (substantia nigra and locus coeruleus) and adrenal medulla (Kaneda et al., 1990). The multiple forms have been found only in humans, and other mammals showed only a type of TH cDNA which is homologous to human TH type 1. We (Ichikawa et al., 1990) have recently found two types of TH mRNA corresponding to type 1 and type 2 in adrenal gland and brain of two species of primate, *Macaca irus* and *Macaca fuscata*. In contrast, only a single form of TH mRNA was detected in one species of insectivore, *Suncus murinus* and rat. Therefore, the multiplicity of TH mRNA may be primate specific.

Although the physiological and pathological significance of the human TH mRNA multiplicity remains to be determined, humans may obtain new regulatory properties of TH by utilizing multiple TH types.

Phosphorylation of TH is one of the most important mechanisms regulating its activity (review by Zigmond et al., 1989). Many kinases are involved in phosphsrylation of TH, i.e., cyclic AMP-dependent protein kinase and Ca^{2+}-calmodulin-dependent protein kinase II (Campbell et al., 1986). The N- and C-terminal regions of TH contain the regulatory and catalytic domains (Ledley et al., 1985). The regulatory domain is modified by phosphorylation. Cyclic AMP-dependent protein kinase mainly phosphorylates Ser-40, and Ca^{2+}-calmodulin-dependent protein kinase II phosphorylates both Ser-19 and Ser-40 (Campbell et al., 1986). These Ser sequences are conserved between rat TH and type 1 human TH. Twelve base pairs insertion of type 2 and type 4 human TH mRNAs may generate a putative phosphorylation site which may alter TH regulation.

Changes in TH in the striatum of MPTP-treated mice

MPTP produces complex biochemical changes in the nigrostriatal DA neurons at early and late stages of MPTP administration in mice. Peripherally administered MPTP is rapidly transported from the blood into the brain through the blood-brain barrier and is oxidized to 1-methyl-4-phenylpyridinium ion (MPP^+), probably by monoamine oxidase (MAO) type B in glial cells. MPP^+ formed in the brain is then specifically taken up into the nigrostriatal DA neurons via DA transporter.

MPP^+ derived from MPTP affects the TH system in a complex way depending upon the length of time after MPTP administration (Hirata and Nagatsu, 1986).

We showed that MPTP, when systemically administered to mice, acutely inhibited tyrosine hydroxylation in situ in striatal tissue slices but not TH activity (Vmax) in vitro in the presence of saturating concentrations of a tetrahydropteridine cofactor, tyrosine, and oxygen (Hirata and Nagatsu, 1986). Acute inhibition of DOPA formation in vivo in the striatum after systemic administration of MPTP to mice was also proved by Pileblad et al. (1985). This acute inhibition of TH may be due to inhibition of phosphorylation of TH. We proved that MPP^+ inhibits both in situ TH activity and TH phosphorylation in rat pheochromocytoma PC 12h clonal cells (Kiuchi et al., 1988). We have also shown that continuous exposure of the striatal DA neurons in rats to MPP^+ in vivo for three hours using a brain microdialysis technique, leads to inactivation of TH without reducing the TH protein content (Ozaki et al., 1988). This inactivation of TH may be due to continued inhibition of phosphorylation of TH by MPP^+, resulting in the formation of inactive, dephosphorylated TH.

At a later stage, after repeated systemic administration of MPTP to mice, both TH protein assayed by a sandwich enzyme immunoassay and TH activity (Vmax) were found to decrease in parallel, and homospecific activity (TH activity/TH protein) did not change (Mogi et al., 1987).

These results suggest that at a later stage after repeated systemic administration of MPTP to mice, TH activity and TH protein in the striatum decrease in parallel, probably owing to cell loss of the nigrostriatal DA neurons. On the 9th days after eight injection of MPTP (30 mg/kg, subcutaneously per day) to C57 BL/6N mice, TH, DOPA decarboxylase, and DA in the striatum were decreased in parallel, but TH, DOPA decarboxylase, DA β-hydroxylase, DA, and noradrenaline in the hypothalamus did not change significanthy. These results may suggest that repeated MPTP administration to mice, at least under these experimental conditions, decreases TH specifically in the nigrostriatal DA regions (Mogi et al., 1988a).

Changes in TH in parkinsonian brains

In collaboration with Narabayashi and coworkers, we examined the changes in catecholamine-synthesizing enzymes and the biopterin cofactor in the parkinsonian human brains. TH activity and the biopterin cofactor, as well as the activities of DA β-hydroxylase and phenylethanolamine N-methyltransferase, are decreased in the parkinsonian brains as compared with those in age-matched controls (Nagatsu, 1975; Nagatsu et al., 1977, 1981). These results indicate that in parkinsonian brains, not only TH in the nigrostriatal DA neurons, but also DA β-hydroxylase in the noradrenaline neurons and phenylethanolamine N-methyltransferase in the adrenaline neurons may be decreased.

We examined both TH activity (Vmax) and TH protein content by a sandwich enzyme immunoassay in parkinsonian human brains (Mogi et al., 1988b). Both the TH protein content and TH activity (Vmax) decreased in parallel in the nigrostriatal regions of the parkinsonian brains as compared with those of control brains. However, the decrease of TH protein was more marked than the decrease of TH activity, resulting in significantly increased TH homospecofic activity (TH activity per TH protein) in parkinsonian brains. This suggests a compensatory TH activation following the reduction of TH protein. This finding agrees with the result by Elsworth et al. (1987) that two months after MPTP treatment in monkeys, the ratio of homovanillic acid to DA was elevated in the putamen and caudate nucleus, suggesting a compensatory increase in transmitter in the surviving neurons. As described previously, we showed that TH activity and TH protein decreased in parallel without any significant change in the homospecific activity appearing on the 9th day, following repeated daily administration of MPTP (30 mg/kg, subcutaneously) to mice for 8 days (Mogi et al., 1987). It is conceivable, therefore, that such compensatory increase in the homospecific activity of TH in the striatum in Parkinson's disease may also occur in primates in some later stage of MPTP-induced parkinsonism. As described above, we have recently shown that the multiple TH mRNAs are observed in humans and monkeys and may be primate specific (Ichikawa et al., 1990). Different forms of TH may also change the phosphorylation of the TH molecules to activate TH.

Conclusion

Historical overview and future prospect on TH in Parkinson's disease are presented. Decreased TH activity in the nigrostriatal DA neuron in the parkinsonian brain was found around 1975. This was also confirmed in MPTP parkinsonism in animals.

Table 1. Comparison of changes in tyrosine hydroxylase (TH) in the brain between parkinsonian patients and MPTP-treated mice

TH changes	Parkinson's disease	MPTP-parkinsonism (mouse)
Time cource		
Early	?	TH activity, inhibition
Middle	?	TH activity, inactivation
Late	TH protein, decrease	TH protein, decrease
Phosphorylation	?	decrease
Homospecific activity (TH activity/TH protein)	increase	no change
Neuron specificity	relatively specific: nigrostriatal DA neurons and noradrenaline neurons	specific: nigrostriatal DA neurons

The decrease in TH activity was found to be in parallel with decrease in TH protein, and it is assumed to be a result of the DA cell loss.

Table 1 summarizes changes in TH activity and TH content between Parkinson's disease and MPTP-induced parkinsonism in mice. Two differences are noted. (1) Homospecific activity of TH (TH activity/TH protein) is increased in Parkinson's disease, but does not change in MPTP-treated mice. (2) The neuronal damage in Parkinson's disease is relatively specific for nigrostriatal DA neurons but is also observed in noradrenaline neurons. In MPTP-treated mice, the damage is localized mainly in the nigrostriatal DA neurons, but not in the noradrenaline neurons.

These differences may be noted in relation to the multiple forms of human TH in contrast to a single form of rodent TH, and to probable different regulatory mechanisms in gene expression and activity regulation of human TH. Since the structure of TH genes and TH proteins may be different between primates and rodents, the molecular changes of TH in Parkinson's disease would be better estimated in MPTP-induced parkinsonism of monkeys. Molecular biological approach to TH in Parkinson's disease is expected to be useful.

Acknowledgment

We are grateful to the support by a Grant-in-Aid for Scientific Research on Primary Areas "Molecular Biology of the Motor System", Ministry of Education, Science and Culture, Japan.

References

Bernheimer H, Birkmayer W, Hornykiewcz O, Jellinger K, Seitelberger F (1973) Brain dopamine and the syndromes of Parkinson and Huntington. Clinical, morphological and neurochemical correlations. J Neurol Sci 20:415–455

Brenneman AR, Kaufman S (1964) The role of tetrahydropteridine in the enzymatic conversion of tyrosine to 3,4-dihydroxyphenylalanine. Biochem Biophys Res Commun 17:177–183

Burns RS, Chiueh CC, Markey SP, Ebert MH, Jacobowitz DM, Kopin IJ (1983) A primate model of parkinsonism: selective destruction of dopamine neurons in the pars compacta of the substantia nigra by N-methyl-4-phenyl-1,2,3,6-tetrahydropyridine. Proc Natl Acad Sci USA 80:4546–4550

Campbell DG, Hardie DG, Vulliet PR (1986) Identification of four phosphorylation sites in the N-terminal region of tyrosine hydroxylase. J Biol Chem 261:10489–10492

Davis GC, Williams AC, Markey SP, Ebert MH, Caine ED, Reichert CM, Kopin IJ (1979) Chronic parkinsonism secondary to intravenous injection of meperidine analogues. Psychiatry Res 1:249–254

Elsworth JD, Deutch AY, Redmond Jr DE, Sladek Jr JR, Roth RH (1987) Effects of 1-methyl-4-phenyl-1,2,3,6-tetrahydropyridine (MPTP) on catecholamines and metabolites in primate brain and CSF. Brain Res 415:293–299

Grima B, Lamouroux A, Boni C, Julien JF, Javoy-Agid F, Mallet J (1987) A single human gene encoding multiple tyrosine hydroxylases with different predicted functional characteristics. Nature 326:707–711

Heikkila RE, Hess A, Duvoisin RC (1984) Dopaminergic neurotoxicity of 1-methyl-4-phenyl-1,2,3,6-tetrahydropyridine in mice. Science 224:1451–1453

Hirata Y, Nagatsu T (1986) Early and late effects of systematically administered 1-methyl-4-phenyl-1,2,3,6-tetrahydropyridine (MPTP) on tyrosine hydroxylase activity in vitro and on tyrosine hydroxylation in tissue slices of mouse striatum. Neurosci Lett 68:245–248

Ichikawa S, Ichinose H, Nagatsu T (1990) Multiple mRNAs of monkeys tyrosine hydroxylase. Biochem Biophys Res Commun 173:1331–1336

Ishii A, Kiuchi K, Matsuyama M, Satake T, Nagatsu T (1990) Ferrous ion activates the less active form of human adrenal tyrosine hydroxylase. Neurochem Int 16:59–64

Kaneda N, Kobayashi K, Ichinose H, Kishi F, Nakasawa A, Kurosawa Y, Fujita K, Nagatsu T (1987) Isolation of novel cDNA clone for human tyrosine hydroxylase: alternative RNA splicing produces four kinds of mRNA from a single gene. Biochem Biophys Res Commun 146:971–975

Kaneda N, Kobayashi K, Ichinose H, Sasaoka T, Ishii A, Kiuchi K, Kurosawa Y, Fujita K, Nagatsu T (1990) Molecular biological approaches to catecholamine neurotransmitters and brain aging. In: Nagatsu T, Hayaishi O (eds) Aging of the brain: cellular and molecular aspects of brain aging and Alzheimer's disease. Japan Scientific Societies Press, Tokyo, and Karger, Basel, pp 53–66

Kiuchi K, Hagihara M, Hidaka H, Nagatsu T (1988) Effect of 1-methyl-4-phenylpyridinium ion on tyrosine hydroxylase in rat pheochromocytoma PC 12h cells. Neurosci Lett 89:209–215

Kobayashi K, Kaneda N, Ichinose H, Kishi F, Nakazawa A, Kurosawa Y, Fujita K, Nagatsu T (1987) Isolation of a full-length cDNA clone encoding human tyrosine hydroxylase type 3. Nucl Acids Res 15:6733

Kobayashi K, Kaneda N, Ichinose H, Kishi F, Nakazawa A, Kurosawa Y, Fujita K, Nagatsu T (1988a) Structure of the human tyrosine hydroxylase gene: alternative splicing from a single gene accounts for generation of four mRNA types. J Biochem 103:907–912

Kobayashi K, Kiuchi K, Ishii A, Kaneda N, Kurosawa Y, Fujita K, Nagatsu T (1988b) Expression of four types of human tyrosine hydroxylase in COS cells. FEBS Lett 238:431–434

Kojima K, Mogi M, Oka K, Nagatsu T (1984) Purification and immunochemical characterization of human adrenal tyrosine hydroxylase. Neurochem Int 6:475–480

Langston JW, Ballard P, Tetrud JW, Irwin I (1983) Chronic parkinsonism in humans due to product of meperidine-analog synthesis. Science 219:979–980

Ledley FD, DiLella AG, Kwok SCM, Woo SLC (1985) Homology between phenylalanine and tyrosine hydroxylase reveals common structural and functional domains. Biochemistry 24:3389–3394

Levitt M, Spector S, Sjoerdsma A, Udenfriend S (1965) Elucidation of the rate-limiting step in norepinephrine biosynthesis in the perfused guinea-pig heart. J Pharmacol Exp Ther 148:1–8

Lloyd KG, Davidson L, Hornykiewicz O (1975) The neurochemistry of Parkinson's disease: effect of L-DOPA therapy. J Pharmacol Exp Ther 195:453–464

Matsuura S, Sugimoto T, Murata S, Sugawara Y, Iwasaki H (1985) Stereochemistry of biopterin cofactor and facile methods for the determination of the stereochemistry of a biologically active 5,6,7,8-tetrahydropterin. J Biochem 98:1341–1348

McGeer PL, McGeer EG (1976) Enzymes associated with the metabolism of catecholamine, acetylcholine and GABA in human controls and patents with Parkinson's disease. J Neurochem 26:65–76

Mogi M, Harada M, Kojima K, Kiuchi K, Nagatsu I, Nagatsu T (1987) Effects of repeated systemic administration of 1-methyl-4-phenyl-1,2,3,6-tetrahydropyridine (MPTP) on striatal tyrosine hydroxylase activity in vitro and tyrosine hydroxylase content. Neurosci Lett 80:213–218

Mogi M, Harada M, Kojima K, Kiuchi K, Nagatsu T (1988a) Effects of systemic administration of 1-methyl-4-phenyl-1,2,3,6,-tetrahydropyridine to mice on tyrosine hydroxylase, L-3,4-dihydroxyphenylalanine decarboxylase, dopamine β-hydroxylase, and monoamine oxidase activities in the striatum and hypothalamus. J Neurochem 50:1053–1056

Mogi M, Harada M, Kiuchi K, Kojima K, Kondo T, Narabayashi H, Rausch D, Riederer P, Jellinger K, Nagatsu T (1988b) Homospecific activity (activity per enzyme protein) of tyrosine hydroxylase increases in parkinsonian brain. J Neural Transm 72:77–81

Mogi M, Kojima K, Harada M, Nagatsu T (1986) Purification and immunological properties of tyrosine hydroxylase in human brain. Neurochem Int 8:423–428

Mogi M, Kojima K, Nagatsu T (1984) Detection of inactive or less active forms of tyrosine hydroxylase in human brain and adrenals by a sandwich enzyme immunoassay. Anal Biochem 138:125–132

Nagatsu T (1975) Biosynthesis and metabolism of dopamine (in Japanese). No-to-Shinkei (Brain and Nerve) 27:1249–1260

Nagatsu T, Kato T, Numata (Sudo) Y, Ikuta K, Sano M, Nagatsu I, Kondo Y, Inagaki S, Iizuka R, Hori A, Narabayashi H (1977) Phenylethanolamine N-methyltransferase and other enzymes of catecholamine metabolism in human brain. Clin Chim Acta 75:221–232

Nagatsu T, Levitt M, Udenfriend S (1964) Tyrosine hydroxylase. The initial step in norepinephrine biosynthesis. J Biol Chem 239:2910–2917

Nagatsu T, Mizutani K, Sudo Y, Nagatsu I (1972) Tyrosine hydroxylase in human adrenal glands and human pheochromocytoma. Clin Chim Acta 39:417–424

Nagatsu T, Yamaguchi T, Kato T, Sugimoto T, Matsuura S, Akino M, Nagatsu I, Iizuka R, Narabayashi H (1981) Biopterin in human brain and urine from controls and parkinsonian patients: application of a new radioimmunoassay. Clin Chim Acta 109:305–311

O'Malley KL, Anhalt MJ, Martin BM, Kalsoe JR, Winfield SL, Ginns EI (1987) Isolation and characterization of the human tyrosine hydroxylase gene: identification of 5'-alternative splice sites responsible for multiple mRNAs. Biochemistry 26: 6910–6914

Ozaki N, Nakahara D, Mogi M, Harada M, Kiuchi K, Kaneda N, Miura Y, Kasahara Y, Nagatsu T (1988) Inactivation of tyrosine hydroxylase in rat striatum by 1-methyl-4-phenylpyridinium ion (MPP$^+$). Neurosci Lett 85:228–232

Pileblad E, Fornstedt B, Clark D, Carlsson A (1985) Acute effects of 1-methyl-4-phenyl-1,2,3,6-tetrahydropyridine on dopamine metabolism in mouse and rat striatum. J Pharm Pharmacol 37:707–712

Rush RA, Kindler SH, Udenfriend S (1974) Homospecific activity, an immunological index of enzyme homogeneity; changes during the purification of dopamine β-hydroxylase. Biochem Biophys Res Commun 61:38–44

Togari A, Kano H, Oka K, Nagatsu T (1983) Simultaneous simple purification of tyrosine hydroxylase and dihydropteridine reductase. Anal Biochem 132:183-189

Zigmond RE, Schwarzchild MA, Rittenhouse AR (1989) Acute regulation of tyrosine hydroxylase by nerve activity and by neurotransmitters via phosphorylation. Ann Rev Neurosci 12:415–461

Correspondence: Prof. Dr. T. Nagatsu, Department of Biochemistry, Nagoya University School of Medicine, 65 Tsurumai-cho, Showa-ku, Nagoya 466, Japan.

Discovery of dopamine deficiency and the possibility of dopa therapy in Parkinsonism

T. Nakajima

Department of Neuropsychiatry, Kyoto Prefectural University of Medicine, Kyoto, Japan

Summary

The particulars regarding the discovery of dopamine deficency in Parkinson's disease and its neurochemical background are described historically. Furthermore, the novel idea of dopa administration for treatment of the disease and syndrome and its first clinical trial are also introduced.

Introduction

"Shaking palsy" (paralysis agitans) was first described in 1817 by the London physician James Parkinson (1817). He concluded his essay noting that "an important object is the leading of the attention of those who, humanely employ anatomical examination in detecting the causes and nature of diseases, particularly to this malady. By their benevolent labours its real nature may be ascertained, and appropriate modes of relief, or even of cure, pointed out." From the beginning of this century, intensive neuropathological studies on this disease or syndrome were performed and showed that the disease derives from degenerative changes in the substantia nigra, a portion of the brain that has extensive connections with the striatum, and is the most prevalent disorder of the extrapyramidal system. However, until about 1950, the functions of the basal ganglia had yet to be clarified, and Greenfield (1955), an authority on neuropathology, was obliged in 1955 to close his review on the pathology of Parkinson's disease with words to the effect that the cause of the disorder could not be detected by anatomical or pathological studies, and that its elucidation must be left to enzyme chemistry or other areas of new research.

L-Dopa therapy in the current treatment of Parkinson's disease or Parkinsonism was scientifically designed on the basis of neurochemical findings, i.e., the regional distribution of dopamine (DA) in the brain and a deficiency of this amine in the disease or syndrome, although many drug therapies in the neuropsychiatric field have been accidentially or empirically composed.

This paper gives the historical particulars of the discovery of DA deficiency in the brain of a patient with Parkinson's disease, and suggests a method of replenishing DA with dopa for the treatment of Parkinsonism.

The neurochemical background of discovery of DA deficiency in Parkinsonism

The examination of distribution of catechol compounds in the human brain (Sano et al., 1959) provided the clue for I. Sano to consider rectification of the abnormality in Parkinsonism with dopa. The researchers involved in neurochemistry in those days focused their attention on catecholamines and endeavored to develop chemical microdeterminations and to clarify the physiological siginificance of catecholamines in the central nervous system (CNS).

In 1946, von Euler (1946) first reported the finding of noradrenaline (NA) in a calf brain. However, the question of whether the origin of NA was vascular or parenchymatous remained. Conclusive evidence was brought forward by Vogt (1954), who demonstrated in 1954 that the amine distributes in high concentrations from the dorsal portion of the brain stem to the hypothalmus, indicating a clear difference from the vascularity in the brain and proving the parenchymatous origin of NA. Furthermore, she suggested that it may be involved in functions of the reticular formation of the brain stem because it is present mainly in this portion. In 1958, Carlsson et al. (1958, 1959) proved that DA exists in the cat brain by using the chemical determination of catechol compounds and its specific distribution in the nuclei of the extrapyramidal system, giving a novel significance to this amine in the CNS, which had hitherto been thought to be just the precursor of NA.

At that time, it was very difficult to measure the catecholamines in biological samples because their concentrations are very low in these samples, which include brain tissues, and because of the sensitivity of the determinations. Therefore, development of sensitive chemical microdeterminations was an interesting subject of research in those days. K. Sano, a young brother of I. Sano, developed a sensitive analytical determination of catechol compounds in human urine, at the Department of Pharmacology, Osaka University Medical School. On the other hand, I. Sano et al. (1959) succeeded in determining catechol compounds in biological samples using alumina absorption of these compounds, their chromatographic separation with Amberlite CG-50 and a combination of two fluorescent reactions, and analyzing NA, DA and dopa in 40 portions of human brain. DA was proven to distribute in high concentrations in the subcortical nuclei of the extrapyramidal system (n. caudatus, globus pallidus, putamen, substantia nigra and n. ruber), suggesting that it is involved in the functions of the extrapyramidal system.

Determination of DA in the brain of a patient with Parkinson's disease

On the 5th of August, 1959, I. Sano (1959) gave a lecture on "Biochemical studies of aromatic monoamines in the brain" at the 15th General Assembly of the Japan Medical Congress held in Tokyo, and reported the results of studies carried out in his department as well as his idea of the pathogenesis of Parkinson's disease, i.e., that monoamines in the CNS, especially DA, may be involved in this malady. His introductory remarks were "as the aromatic monoamines cannot pass through the blood-brain barrier, well-designed methods are needed for an increase or a decrease of these amines in the CNS. To increase the amines, a peripheral administration of their precursor amino acids or the inhibitors of monoamine oxidase (MAO), which is the metabolizing enzyme of monoamines, is used; to decrease the amines, administration of their releasers is used." The gist of his subsequent remarks was as follows.

1) It was confirmed that when reserpine was intraperitoneally injected into a guinea pig, the NA and DA contents in the brain were decreased (these findings were already reported M. Holzbauer and A. Shore, respectively, and B. Brodie demonstrated a decrease in the content of brain serotonin with the injection of the releaser). Furthermore, the animal injected with reserpine became lethargic, irresponsive to external stimuli, and hypokinetic.

2) It was well known that severe Parkinsonism appears in the case of the administration of a large amount of reserpine to psychotic patients. It seemed improbable that this drug-induced Parkinsonism derived from the decrease of serotonin in the CNS, because this amine distributes specifically in high concentrations in the limbic system of the cerebrum, which was hardly thought to be involved in the appearance of Parkinsonism. The study on the distribution of catechol compounds in the human brain revealed the selective occurrence of DA in the subcortical nuclei of the extrapyramidal system, and the reserpine-induced Parkinsonism was possibly implicated in the dysfunction of the extra-pyramidal system derived from DA deficiency, because reserpine reduces the amount of DA in the CNS, especially in the nuclei of the extrapyramidal system.

3) It was confirmed that the intraperitoneal injection of MAO inhibitor, iproniazid or pheniprazine, increased the amounts of NA and DA in the CNS (A. Pletcher et al., already reported an increase of the catecholamines and serotonin in the CNS with MAO inhibitors).

4) In the comparison of the effects of intracarotic injections of NA, DA or dopa into a guinea pig, an increase of DA in the CNS was observed only in the case of the injection of dopa. Furthermore, the injection of dopa induced a recovery in the DA content reduced with reserpine and even a 7-8-fold increase compared to the DA content in the brain of an untreated animal. The animal also recovered rapidly from the lethargic state caused by the releaser, becoming hyperkinetic and hypersensitive.

5) The intraperitoneal injection of tyrosine, the precursor of dopa, into a guinea pig only increased the amount of DA in the CNS slightly.

In the above lecture, the possibility of dopa therapy against Parkinsonism could be inferred clearly, and the thought on the pathogenesis of Parkinson's desease, i.e., that a decrease in DA formation in the CNS may cause this malady, was reported in the review entitled "Die Katechinamine im Zentral-nervensystem" in Klin. Wschr., Vol. 38, issued on the 15th of January, 1960 (Sano, 1960). This pathogenetic hypothesis was verified using a postmortem study of Parkinson's disease in the summer of 1959.

In the beginning of 1959, an untreated patient with Parkinson's disease was transferred to Osaka University Hospital from a mountain village in Wakayama Prefecture; he died 5 months after admission. His brain was used for the analysis of catechol compounds, and a reduced amount of DA was observed in the n. caudatus, putamen, globus pallidus and substantia nigra of the brain, just as expected. This case is the first analysis of catechol compounds in the brain of a patient with Parkinson's disease, and the finding was reported in a special lecture given I. Sano at the first meeting of the Japanese Neuropathological Association, held in Tokyo on the 6th of February, 1960.

Trial of dopa administration for treatment of Parkinsonism

Although I. Sano thought that the finding of DA deficiency in the above case should be confirmed using untreated fresh cases, he tried to administer dopa to 5 patients with Parkinsonism in accordance with his belief that "Parkinsonism must be relieved with the administration of dopa, the precursor of DA." Dopa (200 mg of DL-form) in saline was intravenously injected into a patient pretreated with or without pheniprazine, a MAO inhibitor. The first case was a 65-year-old man whose muscular rigidity was relieved 15–30 min after the injection of dopa. This relief was ascertained electromyographically, and the patient gave the impression of rapid relief. However, this effect was temporary and lasted only for 20–30 min. The same was observed in the other 4 cases.

At that time, dopa was very expensive and available only in the DL-form in Japan. Therefore, I. Sano concluded his lecture with words that "although the results of this clinical trial were very interesting scientifically, the administration of dopa is now impractical because low-priced anti-Parkinson drugs such as trihexyphenidyl are commercially available. Further investigation will be needed for its clinical application." The contents of his lecture were published in Shinkei Kenkyu no Shinpo, Vol. 5, 1960 (Sano, 1960).

Conclusions

At the end of 1960, Ehringer and Hornykiewicz (1960) reported a deficiency of DA as well as NA in the n. caudatus, putamen and globus pallidus of patients with Parkinsonism. Birkmayer and Hornykiewicz (1961) published "Der L-3,4-Dioxyphenylalanin (= DOPA)-Effekt bei der Parkinson-Akinese" in Wien. Klin. Wschr., Vol. 73, issued on the 10th of November, 1961. I. Sano wondered at their optimistic conclusion that Parkinson's disease could be cured with dopa, although almost the same type of the medication as he used, and he immediately traveled to Vienna to meet with Hornykiewicz to exchange views. In those days, I. Sano was in his early thirties, a young and promising associate professor of the Department of Neuropsychiatry, Osaka University Medical School.

References

Birkmayer WZ, Hornykiewicz O (1960) Der 3,4-L-Dioxyphenylalanin (= DOPA)-Effekt bei der Parkinson-Akinese. Wien Klin Wschr 73:787–788

Carlsson A, Lindvist M, Magnusson T, Waldeck B (1958) On the presence of 3-hydroxytyramine in brain. Science 127:471

Carlsson A (1959) The occurrence, distribution and physiological role of catecholamines in the nervous system. Pharmacol Rev 11:490–493

Ehringer H, Hornykiewicz O (1960) Verteilung von Noradrenalin und Dopamin (3-Hydroxytyramin) im Gehirn des Menschen und ihr Verhalten bei Erkrankungen des extrapyramidalen Systems. Klin Wschr 38:1236–1240

Greenfield JG (1955) James Parkinson 1755–1824. In: Critchley M (ed) A bicentenary volume of papers dealing with Parkinson's disease. Macmillan, New York, p 219

Parkinson J (1817) An essay of the shaking palsy. Whittingham and Rowland, London

Sano I (1959) Biochemical studies of aromatic monoamines in the brain. In: Japanese Medicine in 1959. The report of scientific meetings in the 15th General Assembly of the Japan Medical Congress, vol V, pp 607–615

Sano I, Kakimoto Y, Taniguchi K, Takesada M, Gamo J, Nishinuma K (1959) Distribution of catechol compounds in human brain. Biochim Biophys Acta 32: 586–587

Sano I (1960) Biochemistry of extrapyramidal motor system. Shinkei Kenkyu no Shinpo (Adv Neurol Sci) 5:42–48

Sano I, Taniguchi K, Gamo T, Takesada M, Kakimoto Y (1960) Die Katechinamine im Zentralnervensystem. Klin Wschr 38:57–62

Vogt M (1954) The concentration of sympathin in different parts of the central nervous system under normal conditions and after the administration of drugs. J Physiol 123:451–458

Von Euler US (1946) A specitic sympathomimetic ergone in adrenergic nerve fibers (sympathin) and its relations to adrenaline and noradrenaline. Acta Physiol Scand 12:73–97

Correspondence: Dr. T. Nakajima, Department of Neuropsychiatry, Kyoto Prefectural University of Medicine, Kawaramachi-Hirokoji, Kamigyoku, Kyoto 602, Japan.

Genes of human catecholamine-synthesizing enzymes

T. Nagatsu[1], N. Kaneda[1], K. Kobayashi[1], H. Ichinose[1], T. Sasaoka[1],
A. Ishii[1], C. Sumi[1], K. Kiuchi[2], K. Fujita[3], and Y. Kurosawa[3]

[1] Department of Biochemistry, [2] Radioisotope Center Medical Division, Nagoya University School
of Medicine, Nagoya, and [3] Institute for Comprehensive Medical Science, School of Medicine,
Fujita Health University, Toyoake, Aichi, Japan

Summary

We cloned full-length cDNAs and genomic DNAs of human catocholamine-synthesizing enzymes, i.e., tyrosine hydroxylase (TH), aromatic L-amino acid decarboxylase (AADC), dopamine β-hydroxylase (DBH), and phenythanolamine N-methyltransferase (PNMT), and determined the nucleotide sequences and the deduced amino acid sequences. Multiple mRNAs of human TH, human DBH, and human PNMT were discovered by cDNA cloning. Four types of human TH mRNAs are produced by althernative splicing mechanisms from a single gene. The multiple forms of human TH may give additional regulation to the human enzyme. We have succeeded in expressing human TH gene in transgnic mice. The 5'-flanking regions of the genes of human TH, DBH and PNMT contain possible transcription regulatory elements such as cyclic AMP responsive element (CRE) (TH, DBH, PNMT), glucocordicoid responsive element (GRE) (DBH, PNMT), and $Sp1$ (TH, PNMT) binding site.

Introduction

Since catecholamines play a central role in the pathogenesis of Parkinson's disease, the structures, the properties, and the regulation of catecholamine-synthesizing enzymes are of great importance in basic researches on the disease.

Catecholamines (dopamine, noradrenaline, and adrenaline) are synthesized from tyrosine in the central dopamine, noradrenaline, and adrenaline neurons, the peripheral sympathetic noradrenaline neurons, and the adrenal medulla noradrenaline and adrenaline cells, by the following pathway: tyrosine →3, 4-dihydroxyphenylalanine (DOPA) → dopamine → noradrenaline → adrenaline.

Four enzymes are involved in the biosynthesis: (1) tyrosine hydroxylase (TH), (2) aromatic L-amino acid decarboxylase (AADC) or DOPA decarboxylase (DDC), (3) dopamine β-hydroxylase (DBH), and (4) phenylethanolamine N-methyltransferase (PNMT). These four enzymes were purified and characterized from human adrenal medulla or human brain. However, since it was difficult to get enough enzyme proteins from human tissues to determine the complete amino acid sequences, the primary structures of human catecholamine synthesizing enzymes should have been deduced from the nucleotide sequence of the cDNAs.

On the other hand, since 1975, our group has been collaborating with Narabayashi's group (Juntendo University, Tokyo) on biochemical studies of catecholamine-synthesizing enzymes and the biopterin cofactor in the autopsy human brains. As the results, we have observed severe decreases in TH activity and the biopterin cofactor content and moderate decreases in DBH and PNMT activities. Our results suggest that in parkinsonian brains not only DA neurons but also noradrenaline and adrenaline neurons may be moderately impaired (Nagatsu et al., 1977, 1981, 1984). We also found that TH in parkinsonian brains decreased both in activity and enzyme protein content in parallel, but with an significantly increased homospecific activity (activity/enzyme protein) (Mogi et al., 1988). This suggests that TH in parkinssnian brain may have some molecular alteration from the enzyme in control brains besides the dearease in the enzyme content.

These results indicate that sudies on the molecular structures and the gene expression of human catecholamine-synthesizing enzymes are urgent in the basic studies on Parkinson's disease. We have cloned full-length cDNAs and genomic DNAs of human TH, AADC, DBH, and PNMT.

Human tyrosine hydroxylase (TH) gene

TH catalyzes the first step in the biosynthesis of catecholamine neurotransmitters, the formation of DOPA from tyrosine (Nagatsu et al., 1964). TH is an iron-containing pteridine-dependent monooxygenase and expressed in catecholaminergic neurons in discrete regions of the brain, noradrenergic neurons of sympathetic ganglia and sympathetic nerves, and noradrenaline and adrenaline cells of the adrenal medulla. Its activity is modulated by several factors involving modifications of the enzyme molecule by phosphorylation and regulation of its expression (Dahlstrom et al., 1988; Zigmond et al., 1989). As described above, TH may play an important role in the etiology of Parkinson's disease. TH was purified from various tissues of several species, including bovine adenal medulla (Nagatsu and Oka, 1987), rat adrenals (Fujisawa and Okuno, 1987), and rat pheochromocytoma (Tank and Weiner, 1987). Human TH was purified from adrenals (Mogi et al., 1984; Kojima et al., 1984) and brain (Mogi et al., 1986). TH is a homotetramer requiring a tetrahydropterin cofactor and Fe^{2+}. The molecular weight of the subunit of TH from various sources is about 60 kDa, and each subunit may contain one cataytic site, where the substrates, L-tyrosine and molecular oxygen, and a tetrahydropterin cofactor binds to form the products, L-3, 4-dihydroxyphenylanine (DOPA), H_2O, and a quinonoid dihydropterin. The natural cofactor for TH is *(6R)-L-erythro*-tetrahydrobiopterin (Brenneman and Kaufman, 1964; Matsuura et al., 1985).

```
                                                                              -1
                                                                        ACTGAGCC
                                                                             120
ATGCCCCACCCCGACGGCCACCAGCCACAGGCAAGGGCTTCCCGCAGGCCGTGTCTGAGCTGGACGCCAAGCAGGCAGAGCCATCATGGTAAGAGGCGCAGGGCGCCCCGGGGCCCCAGC
 M  P  T  P  D  A  T  T  P  Q  A  K  G  F  R  R  A  V  S  E  L  D  A  K  Q  A  E  A  I  M  V  R  G  Q  G  A  P  G  P  S
              60                                    180
CTCACAGGCTCTTCCGTGGCCTGGAACTGCAGCCCAGCTGCATCCTACACCCCCAAGGTCCCGGTTCATTGGGCGCAGGCAGAGCCTCATCGAGGACGCCCAAGGAGCGG
 L  T  G  S  W  P  G  T  A  A  P  A  A  S  Y  T  P  T  P  R  S  P  R  F  I  G  R  R  Q  S  L  I  E  D  A  R  K  E  R
                                     300                                                                              360
GAGGCCGGCGGTGGCAGCAGCGGCCGCTGCAGTCCCTGGAGCCCGGGACCCCCTGGAGGCGTGTGGCCTTTGAGGAGAAGGAGGGGAAGGCCGTGCTAAACCTGCTTTCTCCCCGAGG
 E  A  A  V  A  A  A  A  A  V  P  S  E  P  G  D  P  L  E  A  V  A  F  E  E  K  E  G  K  A  V  L  N  L  L  F  S  P  R
                                                  420                                                                  480
GCCACCAAGCCCTCGGCCGCTGTCCGAGCTGTGAAGGTGTTTGAGACGTTTGAAGCCAAAATCCACCATCTAGAGACCCGGCCCGCCCAGAGGCCGCGAGCTGGGGGCCCCCACCTGGAG
 A  T  K  P  S  A  L  S  R  A  V  K  V  F  E  T  F  E  A  K  I  H  H  L  E  T  R  P  A  Q  R  P  R  A  G  G  P  H  L  E
                                            540                                                                              600
TACTTCGTGCGCCTCGAGGTGCGCGGAGGGGACCTGGCCGCCCTGCTCAGTGGTGTGCGCCAGGTGTCAGAGGACGTGCGCAGCCCCGGCGGGCCCAAGGTCCCGTGGTTCCCAAGAAAA
 Y  F  V  R  L  E  V  R  R  G  D  L  A  A  L  L  S  G  V  R  Q  V  S  E  D  V  R  S  P  A  G  P  K  V  P  W  F  P  R  K
                                                  660                                                                  720
GTGTCAGAGCTGGACAAGTGTCATCACCTGGTCACCAAGTTCGACCCTGACCTGGACCACCCGGCTTCTCGGACCAGTGTACCGCCAGCCGCAGGAAGCTGATTGCTGAGATC
 V  S  E  L  D  K  C  H  H  L  V  T  K  F  D  P  D  L  D  H  P  G  F  S  D  Q  V  Y  R  Q  R  R  K  L  I  A  E  I
                                     780                                                                              840
GCCTTCCAGTACAGGCACGGCGACCCGATTCCCCGTGTGGAGTACACCGCCGAGGAGATTGCCACCTGGAAGGAGGTCTACACCACGCTGAAGGGCCTCTACGCCACCCACGCCTGCGGG
 A  F  Q  Y  R  H  G  D  D  P  I  P  R  V  E  Y  T  A  E  E  I  A  T  W  K  E  V  Y  T  T  L  K  G  L  Y  A  T  H  A  C
                                                  900                                                                  960
GAGCACCTGGAGGCCTTTGCTTTGCTGGAGCGCTTCAGCGGCTACCGGGAAGACAATATCCCCCAGCTGGAGGACGTCTCGCCCATGCACTCCCCTGAGCCGGACTGCTGC
 E  H  L  E  A  F  A  L  L  E  R  F  S  G  Y  R  E  D  N  I  P  Q  L  E  D  V  S  R  F  L  K  E  R  T  G  F  Q  L  R  P
                                     1020                                                                             1080
GTGGCCGGGCTGCTTTCGGCCCGTGACTTCCTGGCCTCCCTGGCCTTCCGCGTGTTCCAGTGCACCCAGTATATCCGCCACGCGTCCTCGCCCATGCACTCCCCTGAGCCGGACTGCTGC
 V  A  G  L  L  S  A  R  D  F  L  A  S  L  A  F  R  V  F  Q  C  T  Q  Y  I  R  H  A  S  S  P  M  H  S  P  E  P  D  C
                                                  1140                                                                 1200
CACGAGCTGCTGGGGCACGTGCCCATGCTGGCCGACCGCACCTTCGCGCAGTTCTCGCAGGACATTGGCCTGGCGTCCCTGGGGGCCTCGGATGAGGAAATTGAGAAGCTGTCCACGCTG
 H  E  L  L  G  H  V  P  M  L  A  D  R  T  F  A  Q  F  S  Q  D  I  G  L  A  S  L  G  A  S  D  E  E  I  E  K  L  S  T  L
                                     1260                                                                             1320
TCATGGTTCACCGTGGAGTTCGGGCTGTGTAAGCAGAACGGGGAGGTGAAGGCCTATGGTGCCGGGCTGCTGTCCTCCTACGGGGAGCTCTTGCACTGCCTGTCTGAGGACCTGAGATT
 S  W  F  T  V  E  F  G  L  C  K  Q  N  G  E  V  K  A  Y  G  A  G  L  L  S  S  Y  G  E  L  L  H  C  L  S  E  E  P  E  I
                                                  1380                                                                 1440
CGGGCCTTCGACCCTGAGCCTGCGCGCGTGCAGCCCTACCAAGACCAGACCTACCAGTCAGTCTACTTCGTGTCTGAGAGCTTCAGTGACGCCAAGGACAAGCTCAGGAGCTATGCCTCA
 R  A  F  D  P  P  E  A  A  A  V  Q  P  P  Y  Q  D  Q  T  Y  Q  S  V  Y  F  V  S  E  S  F  S  D  A  K  D  K  L  R  S  Y  A  S
                                     1500                                                                             1560
CGCATCCAGCGCCCCTTCTCCGTGAAGTTCGACCCGTACACGCTGGCCATCGACGGTGCTGGACAGCCCCAGGCCGTGCGGCGCTCCCTGGAGGGTGTCCAGGATGAGCTGGACACCCTT
 R  I  Q  R  P  F  S  V  K  F  D  P  Y  T  L  A  I  D  S  P  Q  A  V  R  R  S  L  E  G  V  Q  D  E  L  D  T  L
                                                  1620                                                                 1680
GCCCATGCGCTGAGTGCCATTGGCTAGGTGCACGGCGTCCTGAGGGCCGTCTCCCAACCTCCCCTGGTCCTGCACTGTCCCGAGCTCAGGCCCTGTGAGGGGCTGGGTCCGGGGTGCC
 A  H  A  L  S  A  I  G  ***
                                     1740                                                                             1800
CCCCATGCCCTCCCTGCCCAGGCTCCCACTGCCCCTGCACCTGCTTCTCAGCGCAACAGCTGTGTGTGCCGTGGTGAGGTTGTGCTGCCTGTGGTGAGGTCGTCTTGGCTCCCCAG
                                     1860
GGTCCTGGGGGCTGCTGCACTGCCCTCCGCCCTTCCCTGACACTGTCTGCCCAATCACCGTCACAATAAAGAAAACTGTGGTCCT(A)n
```

Fig. 1. Nucleotide sequence and deduced amino acid sequence of human TH type 4 cDNA. The 81-bp sequence on the solid line and the 12-bp sequence on the dashed line are deleted in human type 2 cDNA and human type 3 cDNA, respectively. The two sequences corresponding to the 93 bp are deleted in human TH type 1 cDNA

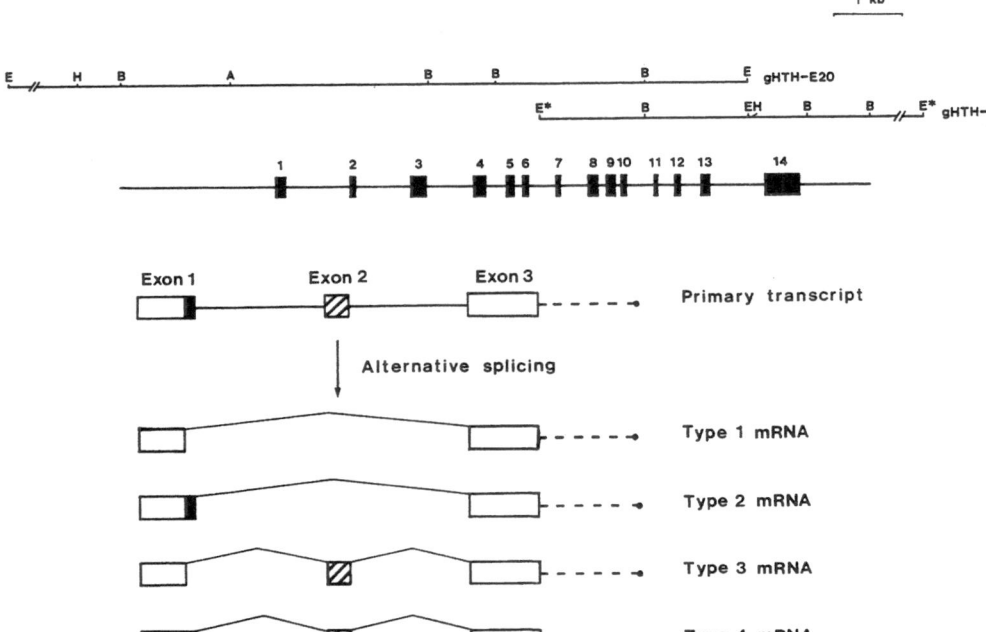

Fig. 2. Structure of the human TH gene, and schematic illustration of the alternative splicing pathway producing the four types of human TH mRNA from the same primrsy transcripts. The 3'-terminal portion of oxon 1, which corresponds to the 12-bp insertion sequence, is indicated by a filled box. The hatched box shows oxon 2 that encodes the 81-bp insertion sequence. The upper shows the retriction enzyme maps of two overlapping genomic clones, gHTH-E20 (~20-kb) and gHTP-1 (~15-kb), Restriction enzyme abreviations: E: *EcoRI*; B: *Bam*HI; H: *Hind*III; A: *Acc*I; E*: *Eco*RI site artificially produced to construct the genomic library (Kobayashi et al., 1988a)

A cDNA containing the entire sequence of rat TH was first isolated from rat pheochromocytoma, and the complete mRNA coding sequence was elucidated (Grima et al., 1985). The open reading frame, including the initation coden, contains 1494 bp that code for 498 amino acids.

Grima et al. (1987) and we (Kaneda et al., 1987; Kobayashi et al., 1987) have found four types of human TH (types 1−4) by cDNA cloning (Fig. 1). Nucleotide sequence analyses of full length cDNAs of type 1 and type 2 (Grima et al., 1987), type 3 (Kobayashi et al., 1987), and type 4 (Kanesa et al., 1987; Le Bourdelles et al., 1988) were completed to deduce the amino acid sequences. These mRNAs are constant for the major part but are distinguishable from one another as to the insertion/deletion of 12- and 81-bp sequences nean the N terminus. The type 2 and 3 mRNAs contain the 12- and 81-bp sequences, repectively, between the 90th and the 91st nucleotides of type 1. In type 4 mRNA, a 93-bp sequence composed of the 12- and 81-bp sequences is inserted at the same position of type 1. Southern blot analysis of human genomic DNA suggested that human TH gene exists as a single gene per haploid DNA, indicating that these different mRNAs are produced through alternative mRNA splicing from a single primary transcript.

O'Malley et al. (1987), La Bourdellès et al. (1988), and we (Kobayashi et al., 1988a) isolated genomic clones encoding the human·TH gene and determined the nucleotide sequence (Fig. 2). The nucleotide sequence of the human TH gene indicated that the human TH gene is composed of 14 exons interrupted by 13 introns and spans approximately 8.5 kb. The exon/intron organization of the human TH gene confirmed our previous prediction that the inserted 81-bp fragment is encoded by an independent exon and that the inserted 12-bp fragment is encoded part of an exon (Kaneda et al., 1987). The 12-bp insertion sequence is derived from the 3'-terminal portion of exon 1 and the 81-bp insertion sequence is encoded by exon 2. The N-terminal region is encoded by the 5'-portion of exon 1, and the remaining region from exon 3 to exon 14 is common to all four kinds of mRNA.

There are two modes of alternative splicing. One is the alternative use of two donor sites in a single exon, exon 1. In Fig. 3A, arrows show the two alternative donor sites in exon 1, the sequence from 117 to 125, ATG $\downarrow$ GTAAGA, and the sequence from 129 to 135, CAG $\downarrow$ GTAGGT, which match 7 and 8, respectively, of 9 nucleotides, with the consensus sequence as a splice donor, ^C_AAG $\downarrow$ GT^A_GAGT. The selection of these potential donor sites determines the insertion/deletion of the 12-bp sequence. The other mode of alternative splicing is the inclusion/exclusion of exon 2 in the spliced products, which determines expression of type 1,2, or type 3,4 mRNA. All of the other 12 exons downstream from exon 3 are spliced and incorporated into mature mRNAs. As shown in Fig. 3B, computer-assisted analysis of the second structure of the primary transcript led to the prediction of four stable hairpin loops in introns 1 and 2: in intron 1 the sequence from 299 to 325 complements that from 1005 to 1029 (structure A, $G = -48.3$ kcal/mol); in intron 2 the sequence from 1197 to 1213 complements that from 1536 to 1553 (structure B, $G = -26.0$ kcal/mol), and the sequence from 1426 to 1443 complements that from 1842 to 1859 (structure C, $G = -38.8$ kcal/mol); and the sequence from 1004 to 1017 in intron 1 pairs with that from 1500 to 1512 in intron 2 (structure D, $G = -37.1$ kcal/mol). Hairpin loops of structures A, B, and C may facilitate the inclusion of exon 2 into mRNA by juxtaposing exons 1 and 2 as well as exons 2 and 3. When the hairpin loop of structure D is formed, the hairpin loop of structure A should be destroyed because the sequence from 1005 to 1016 is common to both structure A and structure D. The hairpin loop of structure D may be involved in the joining of exon 1 derectly to exon 3, preventing the inclusion of exon 2 into mRNA by physically separating exon 2 from exons 1 and 3 through a 14-nucleotide perfect-matched stem structure. We assume the presence of trans-acting factors that stalilize the hairpin structure and presume that they discriminate between structures A and D. When the hairpin loop of structure D should be destroyed for the same reason as described above.

```
5'--- gtctacgagacacacggcctggaatcttctggagaagcaaacaaatgcctcctgacatctgagctggaggctggatccccgtctggggcttctgggtcggtcttgccac        -359

gaggtctggtgttcattaaaagtgtgccctttggctccctggctgtgcttcctggaggctgtggggccaagggaccctggctgtctcagccccgcagagcacg               -239

agccccctggtccccgcaagcccgcgggctgaggatgatcagacagggctggggagtgaaggcaattagattcacgacgagcccttctctgcgctccctcctcctcaccacc      -119

ccgctccatcaggcacgcaggcaggcagggggaagg tggggaacccagaggggccttgacgtcagctcagc [TATAA] gaggctgctgggcaggctgtgg              -1
                                                                                  Exon 1
AGACGGACGCCCGGACCTCCACACTGAGCC ATGCCACCCGACGCGCACCACGCCACAGCGCCAAGGGCTTCCGCAGGGCCGTGTCTGAGCTGGACGTGCAGGCAGGCCAGGCCATCATG   119
                               MetProThrProAspAlaThrThrProGlnAlaLysGlyPheArgArgAlaValSerGluLeuAspValGlnAlaLysGlnAlaGluAlaIleMet

GTAAGAGGGCAG gtaggtgcccggcggccggccagtgagcggagcccaggctggtgccagctgcctcctgctactcccagcctggctggcagcccaggctcaggtccatgcaaac   236
ValArgGlyGln

ccctgggacggcggcggtggatgtggaggccctgggcacgcggcatccccctgcctggctgtttgagtcctgttgggtggagggtgaggtgatgcctgtccctgtgtgtgccccttaggcc   356
                                                                                          (A)

gacctcctcggggtcgtgtggggtcctctgtctgtttcatcctgaatcttaacgatcggaatggtggaacaaaatcatccaaaaatccaagatggccagaggtcccggcctgctgc       476

accagccccacccctactcccacctgccctgccctctctgcccagctgccctagtcagcacccaacagctgccgctgcttggggaggcagcccaaggccttccaggcttctag         596

cagcagctcatggctggggggtcctggccaaatcaaagggtatctgggtctgggtgaatcccttgccctgtcctccctcattccctcatcatcattcatcc                   716

atccattcattcattcaccatggagtctgtgttccctgtgacctgtacagggactgtgtgggcagctgataatcgggagctttcagccacagg                          836

agggtcttcggtgcctccttgggcacttcaggaacctggggctccctggcacatttaaaatgggtttttatttatttatagacttgatttatgaaatgtggtgagttgtagcagtgtcatttcca   956

ggtacctttctcaggaacacaggggccctccccccgtcctcccccgctcccccctaccctcccccacaggctcccatcaccccccccag [GGGGCCCGGGGGCCCAGCCTC]   1073
                                                                                            GlyAlaProGlyProSerLeu

gtaagtaagaggggactctgggagggggcttctgctgctcccctcatgttccacaac                                                           1190

cctggaagtcaggatgaagctgattcttctctacaagggcccagagccttctgggagttcagctccaaggatgagcccagtgtctgcaagtcccccctgtcccagccgctggg       1310

acggctctggatcgagggggtcagaggctcagaggccgagcccaggagagacacctgcgcccagagctatgacaaaggtggagggatgacaaggcagccagagcctggggtggca   1430
                                                                                                            (C)

cagagggggcagggcccggagcaggtgtcctgatggagtgtcctgatccccgtcccccttccctgcggcagccaggaaaggggtggggggtaggggggtgttcactggggcccgtgggggcagctccctcctga   1550
                                                                                                                    (B)

gctgccgtcccccccggcagccgatgccactgtccactgtccatcaagacatcgcccctctcccccccaccactgcttctgactgggtgccccatcggaggaaggggtgcaatgcccgcaggcacctcggct   1670
                                                                                                                    (D)

tcgcccacagagtgggagggacagcggccgagcacctccccagccacctgctctcatccctgctgcatcaagaaacgctgctctcatccctgtggccatccagct            1790

agcatctgcccacagcaggcacacagtaggccctcaaaaacgctgctctcatccctgctgcatcaagaaacgctgctctcatccctgtggcggctctc                   1910
                                                                                          (C)

agcaggtggagaggcatgggtgcccctgtccccacag [TCCCGCGGTTCATTGGGCGCAGCAGCCTCATCGAGGAGCCCGCAAGGAGCGCGGAGCGGGTGGCAGCAGCG]   2027
                                       SerProArgPheIleGlyArgArgGlnSerLeuIleGluGluAspArgLysGluArgGluAlaAlaAlaAla

GCGCTGCAGCTCCCCTCGGAGCCCGGGGCTGGAGGTGGCCGTTTGACGAGGAAGAGGGGAAGGCCGTGTAAACCTGTCTTCTCCCGAGGGCACCAAGCCCTGCGCGCTG   2147
AlaAlaAlaValProSerGluProGlyLeuGluValAlaValAlaPheGluGluGlyLysGluValAlaValLeuAsnLeuLeuPheSerProArgAlaThrLysProSerAlaLeu

TCCCGAGCTGTGAAGTGTTTGAG gtgagctggtggccttctgtccctgggcagtcacctgtggggtcggtgtggggcagtcgtgtggggcgtgtggttctgacccctatagcagaggtgcag   2264
SerArgAlaValLysValPheGlu

ctgccaggcccccgagtcccgacaggatgcagcaggatggcagcaggagtccagcctcagtctgacaccccccatggacctagccacaccccgtgtttttgagggatcc ---3'   2367
```

Exon 2
```
ACAGGCTCTCCGTGGCTGGAACTGCAGCCCCAGCTGCATCCTACACCCCACCCCAAGG
ThrGlySerProTrpProGlyThrAlaAlaProAlaAlaSerTyrThrProThrProArg
(D) (A)
```

Exon 3

(B)

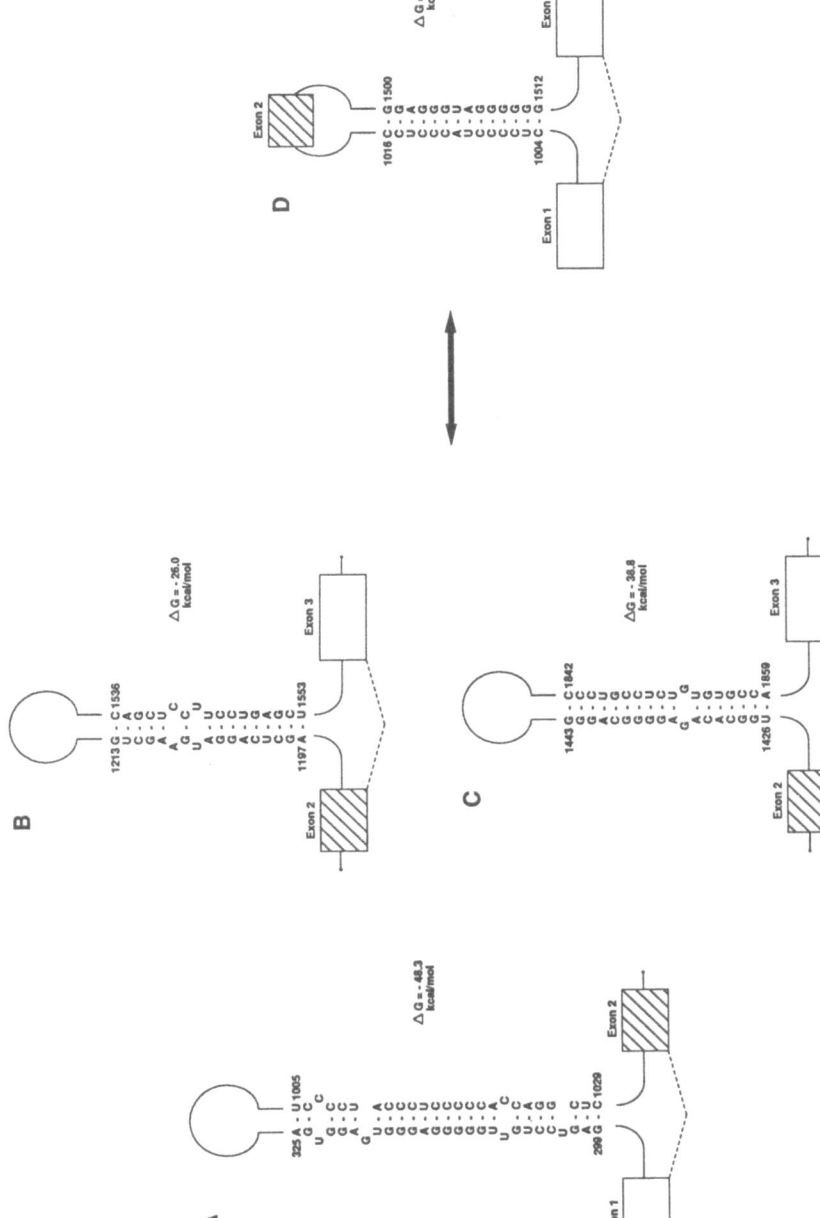

Fig. 3. Nucleotide sequence from upstream of exon 1 to down stream of exon 3 and potential secondary structures in introns 1 and 2 of the primary transcript. (A) Exons 1,2, and 3 are boxed with thin lines. The typical TATA box is boxed in a thick line. Arrows show the alternative donor sites in exon 1. The sequences which complement each other to form the hairpin loops (stractures A, B, C, and D in Fig. 3B) are underlined. (B) The secondary structure was analyzed with the computer program with minimum matching lenghth = 14 bp and matching percentage > 85%. The four kinds of hairpin loops (structures A, B, C, and D) are shown. The nucleotide numbers are given at the start and end points of the hairpin structures. The dushed lines show the possible alternative splicing pathway shown.

(Kobayashi et al., 1988a)

We (Ichikawa et al., 1990) have found two types of TH mRNA corresponding to type 1 and type 2 human TH mRNA in the adrenal gland and brain of two species of monkeys (*Macaca irus* and *Macaca fuscata*). In contrast, only a single form of TH mRNA corresponding to type 1 human TH mRNA was detected in one species of insectivore, *Sunkus murinus*. We suggest that the multiplicity of TH mRNA is primate specific.

Grima et al. (1987) reported the tissue distribution of the mRNAs of human TH by means of S1-protein analysis; the adrenal medulla expressed type 1,2, and 3 mRNAs, whereas the substantia nigra (dopaminergic neurons) and locus coeruleus (noradrenergic neurons) of the brain expressed only type 1 and 2 mRNAs. We (Kaneda et al., 1990) examined the expression pattern of multiple types of human TH mRNA in human brain and adrenal medulla by the primer extension method (Fig. 4A). As shown in Fig. 4B, in adult adrenal medulla relative ratio of type 1:2:3:4 was about 0.9:1.0:0.1:0.3 and in brain (substantia nigra), type 3 was undetectable and the relative ratio of each type was 0.5:1.0:0:0.1 in adult brains. In both tissues, type 1 and 2 were major species. Type 3 was much more expressed in human pheochromocytoma, in which the ratio was about 0.9:1.0:0.3:0.1. We investigated the effect of aging on the expression pattern of TH isozymes. In adrenal gland of neonate (2 days old) or fetus (24 weeks after gestation), types 1 and 2 were predominant and the ratio was 0.7:1.0:0:0.4−0.7. Although type 3 mRNA was not detected, this might be due to the tiny amount of the tissue. In brain stem of these very young tissues, the ratio was about 1.0:1.0:0:0.1. The relative ratio of type 1 TH mRNA appears to be larger than in aged brain. The results may suggest that TH type 1, which exhibited the highest homospecific activity in COS cells (Kobayashi et al., 1988b), may be most susceptible to brain aging. Our results (Kaneda et al., 1990) by the primer extension method agrees with the reports by Grima et al. (1987) and Le Bourdellès et al. (1988) by S1 nuclease mapping in that in human tissues types 2 and 1 are predominant.

We (Kobayashi et al., 1988a) compared the 5′-flanking region of the human TH gene with that of the rat TH gene (Lewis et al., 1987). They show about 70% homology and there are many conserved blocks. One conserved sequence, CCCCCGCCTC, is homologous to the consensus sequence of the binding site of the transcription factor *Sp1* which is known to activate the transcription initiation of many mammalian genes. Another sequence, TGACGTCA, is highly homologous to the conserved sequence of the cyclic AMP-reponsive element (CRE) for transcription activation of various human genes (Kobayashi et al., 1988a; Le Bourdellès et al., 1988; Coker et al., 1988).

We (Kobayashi et al., 1988b) attempted expression of the types 1−4 human TH in COS cells. Expression vectors plasmid pAS-TH 1−4, having human TH type 1−4 cDNAs, respectively, were constructed, and COS cells were transfected with each of pAS-TH 1−4 gave major immunoreactive bands at 61, 61, 65, 65 kDa, respectively, by Western blotting, and showed TH activity. The Km values of the four types for L-tyrosine and (6RS)-methyl-5,6,7,8-tetrahydropterin were similar. The type 1 human TH had the highest homospecific activity (activity/enzyme protein), the values for the other

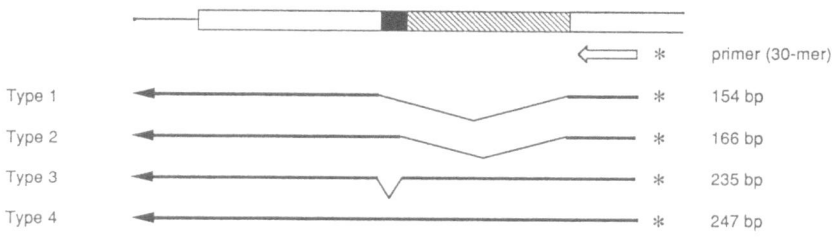

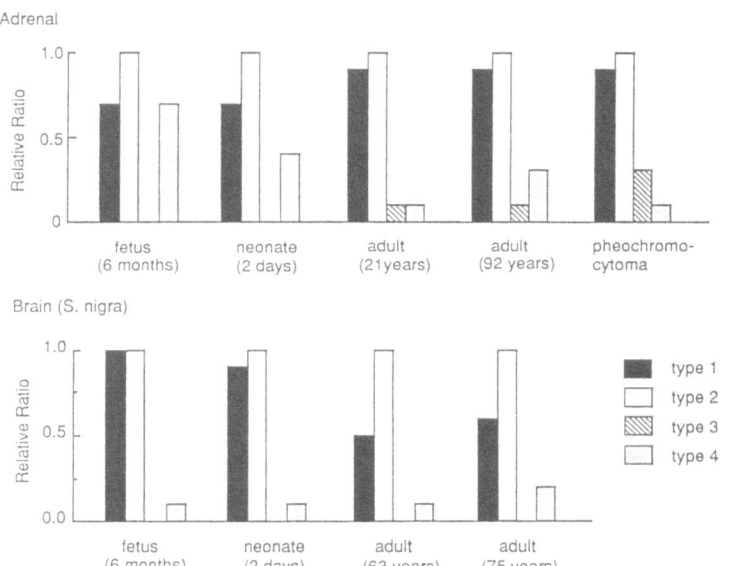

Fig. 4. Primer extension analysis of TH mRNA types in human adrenal and brain. (A) The expected length of the extension products in each type of mRNA. A synthetic oligonucleotide (30-mer) complementary to nucleotide 189–218 in type 4 mRNA was used as the primer. (B) Expression pattern of four types of TH mRNA in human adrenal and brain (substantia nigra). Total RNA (ca. 50 g) extracted from each tissue was subjected to the primer extension analysis, and relative amount of each mRNA type was estimated by densitometry (Kaneda et al., 1990)

enzymes ranging approximately from 0.3 to 0.4. Thus the insertion sequences seem to inhibit TH activity. The human TH type 2 was also expressed in cells using a baculovirus vector, and the results of their kinetic analysis are similar to our data on TH type 2 expressed in COS cells (Ginns et al., 1988). Horellou et al. (1988) injected the full coding sequences of type 1–4 human TH cDNA inserted into the SP6 vector into frog *Xenopus* oocytes, and active TH was produced; the homospecific activity was the highest in the type 1 human TH, and lower in the types 2–4 human TH.

In collaboration with us, Uchida and Kohsaka et al. (1987, 1988, 1989, 1990a, b) have been trying to use genetically manipulated nonneuronal cells as donor cells for intracerebral grafting that produce catecholamines or the precursor, L-DOPA. C6 cells transfected with type 2 TH cDNA were able to express active TH molecules and release L-DOPA from the cells into the culture medium. Since C6 cells lack DOPA decarboxylase, DOPA formed from tyrosine is not decarboxylated to dopamine. Rat kidney fibroblasts, NRK-49F cells transfected with type 2 TH cDNA, which lack the tetrahydrobiopterin synthesis system, expressed active TH and produced DOPA when tetrahydro-biopterin is added exogeneously in culture and in intracerebral grafting. Ishii et al. (1990) showed that the natural (6R)-L-*erythro*-tetrahydrobiopterin is much more active for DOPA formation than the unatural (6S) isomer in culture of NRK-TH type 2 fibroblasts.

We (Kaneda et al., 1990, 1991) have recently succeeded in producing transgenic mice carrying multiple copies of human TH gene. The transgenes were transcribed correctly and expressed specifically in the brain and adrenal gland. Human TH mRNA level in the brain was about 50-fold higher than the endogenous mouse TH mRNA. In situ hybridization demonstrated an enormous region-specific expression of the transgene in substantia nigra and ventral tegmental area. TH-immunoreactivity in these brain regions, though it was not comparable to the increment of the mRNA, was increased definitely in transgenic mice. This observation was also supported by Western blot analysis and TH activity measurements. However, catecholamine levels in transgenics were not significantly different from those in nontransgenics. These results suggest unknown regulatory mechanisms at translational and/or post-translational for human TH gene expression and for the catecholamine levels in transgenic mice.

Human aromatic L amino acid decarboxylase (AADC) gene

AADC (Lovenberg et al., 1962) catalyzes the 2nd step of catecholamine biosynthesis, decarboxylation of L-DOPA in catecholaminergic neurons. AADC is also thought to catalyze decarboxylation of L-5-hydroxytryptophan to sero-tonin in serotonergic neurons. AADC is a homodimer and requires one pyridoxal phosphate molecule per subunit. We (Ichinose et al., 1989) isolated and characterized a full-length cDNA clone encoding human AADC. A human pheochromocytoma cDNA library was screened by use of an oligonucleotide probe corresponding to a partial amino acid sequence of the enzyme purified from the human pheochromocytoma. The isolated cDNA clone encoded a protein of 480 amino acids with a culculated molecular mass of 53891 (Fig. 5). The amino acid sequence, Asn-Phe-Asn-Pro-His-Lys-Trp, around a possible cofactor (pyridoxal phosphate) binding site was identical in human, *Drosophila*, rat, and pig enzymes.

We (Sumi et al., 1990) transfected the full-length cDNA of AADC. The cells expressed AADC as a major immunoreactive band at about 50 kDa. The

```
                                                                                -1
GGAGAGAGAGGAGGACGAGAGAGCAAGTCACTCCCGGCTGCCTTTTCACCTCTGACGAGCCCAGACACC
                                                               60            120
ATGAACGCAAGTGAATTCCGAAGGAGGAGGAAGGAGATGTGATTACGTGGCCAACTACATAGAGGCATTGAGGACGCCAGGTTCACCCTGACGTGGAGCCCGGGTACCTGCGGCCG
 M  N  A  S  E  F  R  R  R  G  K  E  M  V  D  Y  V  A  N  Y  M  E  G  I  E  G  R  Q  V  Y  P  D  V  E  P  P  G  Y  L  R  P
                                                              180            240
CTGATCCCTGCGCGCTGCCCCTCAGGAGCCAGACACGTTGAGGACATCATCAACGACGTTGAGAGAGTAATCATGCCTGGGGTGACGCACTGGCACAGCCCCTACTTCTTCGCCTACTTC
 L  I  P  A  A  A  P  Q  E  P  D  T  F  E  E  D  I  I  N  D  V  E  K  I  I  M  P  G  V  T  H  W  H  S  P  Y  F  F  A  Y  F
                                                              300            360
CCCACTGCCAGCTCGTACCCGGCCATGCTTGCCGACATGCTGTGCGGGCACATTGGCTGCATCGGCTTCTCCTGGGCCGCCAGCATGCCAGCTGCACGAGCTGGAGACTGTGATGATGGAC
 P  T  A  S  S  Y  P  A  M  L  A  D  M  L  C  G  A  I  G  C  I  G  F  S  W  A  A  S  P  A  C  T  E  L  E  T  V  M  M  D
                                                              420            480
TGGCTCGGGAAGATGCTGGAACTACCAAAGGCATTTTTGAATGAGAAAGCTGGAGAAGGGGAGGGAGTGATCCAGGGAAGTGCCAGTGAAGCCACCCTGGTGGCCCTGCTGGCCGCTCGG
 W  L  G  K  M  L  E  L  P  K  A  F  L  N  E  K  A  G  E  G  G  G  V  I  Q  G  S  A  S  E  A  T  L  V  A  L  L  A  A  R
                                                              540            600
ACCAAAGTGATCCATCGGCTGCAGGCAGCGTCCCCAGAGCTCACACAGGCCGCTATCATGGAGAAGCTGGTGGCTTACTCATCCGATCAGGCACACTCCTCAGTGGAAAGAGCTGGGTTA
 T  K  V  I  H  R  L  Q  A  A  S  P  E  L  T  Q  A  A  I  M  E  K  L  V  A  Y  S  S  D  Q  A  H  S  S  V  E  R  A  G  L
                                                              660            720
ATTGGTGGAGTGAAATTAAAAGCCATCCCCTCAGATGGCAACTTCGCCATGCGTGCGTCTGCCCTGCAGGAGGCCCTGGCCTGGCCTGATTCCTTTCTTTATGGTT
 I  G  G  V  K  L  K  A  I  P  S  D  G  N  F  A  M  R  A  S  A  L  Q  E  A  L  E  R  D  K  A  A  G  L  I  P  F  F  M  V
                                                              780            840
GCCACCCTGGGGACCACAACATGCTGCTCCTTTGACAATCTCTTAGAAGTCGGTCCTATCTGCAACAAGGAAGACATATGCGTGCACGTTGATGCAGCCTACGCAGGCAGTGCATCCATC
 A  T  L  G  T  T  C  C  S  F  D  N  L  L  E  V  G  P  I  C  N  K  E  D  I  W  L  H  V  D  A  A  Y  A  G  S  A  F  I
                                                              900            960
TGCCCTGAGTTCCGGCACCTTCTGAATGGAGTGGAGTTGCAGATTCATTCAACTTTAATCCCCACAAATGCTATTGGTGAATTTGACTGTTCTGCCATGTGGGTGAAAAAGAGAACA
 C  P  E  F  R  H  L  L  N  G  V  E  F  A  D  S  F  N  F  N  P  H  K  W  L  L  V  N  F  D  C  S  A  M  W  V  K  K  R  T
                                                             1020           1080
GACTTAACGGGAGCCTTTAGACTGGACCCCACTTACCTGAAGCACAGCCATCAGGATTCAGGGCTTATCACTGACTACCGGCATTGGCAGATACCATCTGGGCAGAAGATTCCGCTCTTTG
 D  L  T  G  A  F  R  L  D  P  T  Y  L  K  H  S  H  Q  D  S  G  L  I  T  D  Y  R  H  W  Q  I  P  L  G  R  R  F  R  S  L
                                                             1140           1200
AAAATGTGGTTTGTATTTAGGATGTATGGAGTCAAAGGCTTGCAGGCTTATATCCGCAAGCATGTCCAGCTGTCCATGAGTTTGAGTCACTGGTCGCCAGGATCCCGCTTTGAAATC
 K  M  W  F  V  F  R  M  Y  G  V  K  G  L  Q  A  Y  I  R  K  H  V  Q  L  S  H  E  F  E  S  L  V  R  Q  D  P  R  F  E  I
                                                             1260           1320
TGTGTGGAAGTCATTCTGGGGCTTGTCTGCTTCCGGCTTAAGGGTTCCAACAAAGTGAATGAAGCTCTTCTGCAAAGAATAAACAGTGCCAAAAAAATCCACTTGGTTCCATGTCACCTC
 C  V  E  V  I  L  G  L  V  C  F  R  L  K  G  S  N  K  V  N  E  A  L  L  Q  R  I  N  S  A  K  K  I  H  L  V  P  C  H  L
                                                             1380           1440
AGGGACAAGTTTGTCCTGCGCTTTGCCATCTGTTCTCGCACGGTGGAATCTGCCCATGTGCAGCGTGCCTGGGAACACATCAAAGAGCTGGCGGCGCTGCTGCGAGCAGAGAGGGAG
 R  D  K  F  V  L  R  F  A  I  C  S  R  T  V  E  S  A  H  V  Q  R  A  W  E  H  I  K  E  L  A  A  D  V  L  R  A  E  R  E
                                                             1500           1560
TAGGAGTGAAGCCAGCTGCCAGGAATCAAAAATTGAAGAGAGATATATCTGAAAACTGGAATAAGAAGCAAATAAAATATCATCCTGCCTTCATGGAACTCAGCTGTCTGTGGCTTCCCATG
 ***
                                                             1620           1680
TCTTTCTCCAAAGCCATCCAGAGGGTTGTGATTTTGTCTGCTTAGTCATCTCATCAACAAGAAATATTATTTGCTAATTAAAAAGTTAATCTTCATGGCCATAGCTTTATTCATTAGCTG
                                                                            1800
TATTTGTGATAAACATATAGATTTCATGTCTGCAGTCATCAGAAGTGGTAGAACTACTGATATATATTTTCCAGGCAATCAATGTTCACGGCAACTTGAAATTATATCTGTGGTCTTCAAAT
                                                             1740
TGTCTTTTGTCATGTGGCTAAATGCCTAATAAACAATTCAAGTG (A) n
```

Fig. 5. Nucleotide sequence of cDNA encoding human AADC and the deduced amino acid sequences (Ichinose et al., 1989)

expressed enzyme catalyzed the decarboxylation of L-DOPA, L-5-hydroxy-tryptophan and L-*threo*-3,4-dihydroxyphenylserine (DOPS). The expressed enzyme had decarboxylase activity in the absence of exogenous pyridoxal phosphate with both L-DOPA and L-5-hydroxytryptophan as substrate. With L-DOPA as substrate, the enzyme had the highest activity within a narrow pH range around 6.5. In contrast, the activity of the enzyme with L-5-hydroxytryptophan as substrate had a broad optimal pH range from 7.0 to 8.4. Addition of pyridoxal phosphate to the incubation mixture greatly enhanced the activity for both L-DOPA and L-5-hydroxytryptophan. This result definitely proves that a single enzyme catalyzes the decarboxylation of both L-DOPA in catecholamine neurons and L-5-hydroxytryptophan in serotonin neurons.

Human dopamine beta-hydroxylase (DBH) gene

DBH catalyzes the third step of catecholamine biosynthesis, i.e., conversion of dopamine to noradrenaline. DBH is a copper-containing, ascorbate-requiring monooxygenase (Friedman and Kaufman, 1965). The enzyme is localized in noradrenaline and adrenaline neurons in the brain and in noradrenaline neurons of periphenal sympathetic nerves and noradrenaline and adrenaline cells of the adrenal medulla. DBH is glycoprotein and a 290 kDa homotetramer consisting of a 75 kDa subunit with 2 atoms of copper per submit. DBH is localized in synaptic vesicles in noradrenaline neurons and in chromaffin granules in adrenomedullary cells. The enzyme is secreted together with noradrenaline or adrenaline into blood or cerebral spinal fluid from central and peripheral noradrenaline neurons, and the enzyme leves in the body fluids change in various diseases (Nagatsu, 1977). Patients suffering from severe orthostatic hypotension were found to be deficient in noradrenaline, adrenaline and DBH, suggesting congenital DBH deficiency (Robertson et al., 1986; Man in't Veld et al., 1987).

Lamouroux et al. (1987) isolated a cDNA clone encoding human DBH and reported the nucleotide sequence and the deduced amino acid sequence. This clone was identical to our type A cDNA (Kobayashi et al., 1989).

We (Kobayashi et al., 1989) isolated two different cDNAs (type A and B) for human DBH and its genomic DNA, and showed that the two mRNAs are generated through alternative polyadenylation from a single gene. Type A mRNA (2.7 kb) and type B mRNA (2.4 kb) encoded the same amino acid sequence and were different only in the 3'-untranslated region. Type A mRNA contained a 3'-extension of 300-bp at the end of the type B mRNA sequence (Fig. 6).

We (Kobayashi et al., 1989) isolated genomic DNA of human DBH (approximately 23 kb). It was composed of 12 exons and 11 introns, and existed as a single gene. Exon 12 encoded the 3'-terminal region of 1013 bp of type A, including the 300-bp sequence. Northen hybridization and S1 nuclease mapping experiments supported the conclusion that alternative use of two polyadenylation sites from a single DBH gene generates two different mRNA

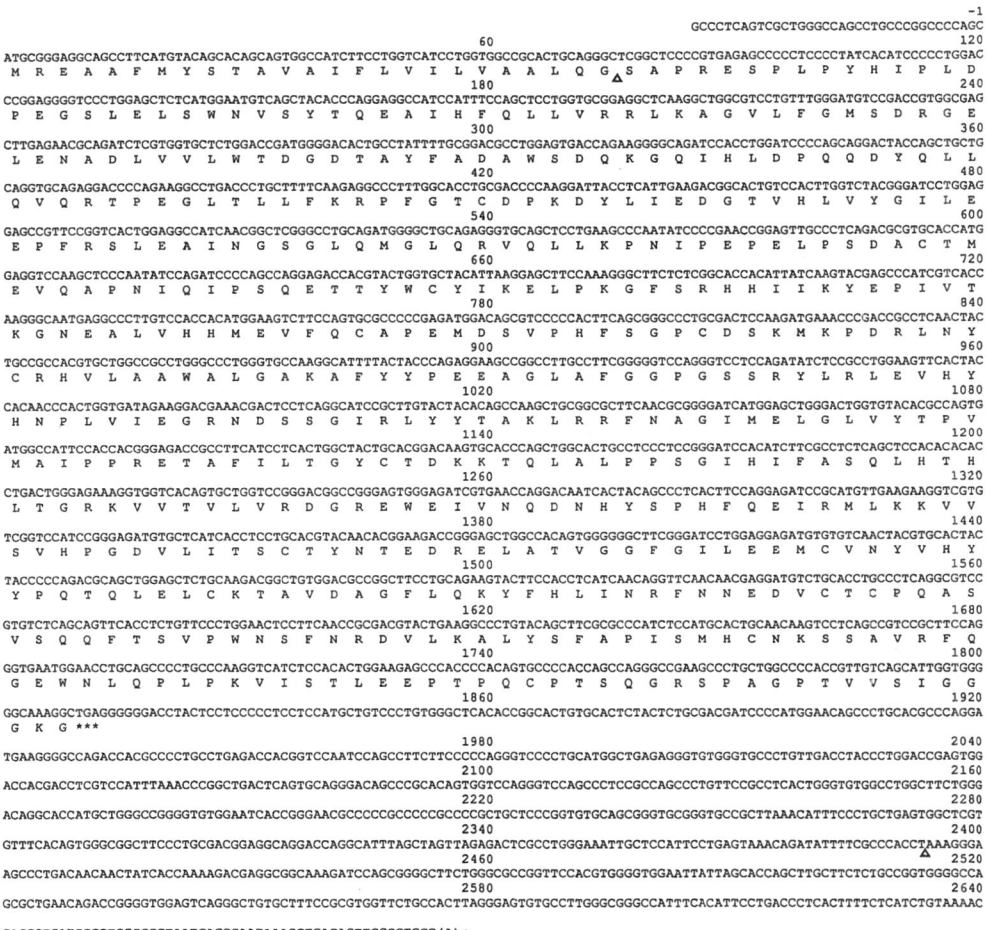

Fig. 6. Nucleotide sequence of cDNA encoding human DBH type A. The arrow between Gly[25] and Ser[26] indicates the signal peptide sequence (Met[1]-Gly[25]) and the arrow after T[2513] indicates the poly A addition site of type B mRNA (Kobayashi et al., 1989)

types, types A and B. The ratio of type A to type B mRNAs in human pheochromocytoma was approximately 1.0 to 0.2. The functional significance of the production of multiple mRNAs having different 3'-untranslated regions through alternative polyadenylation is not known. The 3'-untranslated region may be involved in mRNA stability and translational efficiency.

We (Kobayashi et al., 1989) found possible transcription regulatory elements, TATA, CCAAT, CACCC, GC boxes, and glucocorticoid and cyclic

```
                                                                                                                    -1
                                                                                                              GGCAGC
                                                                                                                 120
ATGAGCGGCGCCAGACCGTAGCCCCAATGCGGGCCAGCCCCTGACTCGGCCCCGGGCCAGGCGGCGGTGGCTTCGGCCTACCAGCGCGCGCCTACCTCCGCAACAACTAC
 M  S  G  A  D  R  S  P  N  A  G  A  A  P  D  S  A  P  G  Q  A  A  V  A  S  A  Y  Q  R  F  E  P  R  A  Y  L  R  N  N  Y
                        60                                                        180                             240
GGCGCCCCTCGCGGGGACCTGTGCAACCCGAACGGCGTCGGGCCGTGGAAGCTGCGCTGCTTGGCGCAGACCTTCGCCACGGTGAAGTGTCCGGAGCACCCTCATCGACATTGGTTCA
 A  P  P  R  G  D  L  C  N  P  N  G  V  G  P  W  K  L  R  C  L  A  Q  T  F  A  T  G  E  V  S  G  R  T  L  I  D  I  G  S
                                  300                                                                              360
GGCCCCACCGTGTACCAGCTGCTCAGTGCCTGCTCAGCCACTTTGAGGACATCACCATGACAGATTTCCTGGAGGTCAACCGCCAGGAGCTGGGCGCTGCTCGAGGAGGAGCCGGGGGCC
 G  P  T  V  Y  Q  L  L  S  A  C  S  H  F  E  D  I  T  M  T  D  F  L  E  V  N  R  Q  E  L  G  R  W  L  Q  E  E  P  G  A
                                                420                                                               480
TTCAACTGGAGCATGTACAGCCAACATGCCTGCCTCATTGAGGGCAAGGGGAATGCTGGCAGGATAAGGAGCGCCAGCTGCCAGCCAGGGTGAAACGGGTCCTGCCATCGACGTGCAC
 F  N  W  S  M  Y  S  Q  H  A  C  L  I  E  G  K  G  E  C  W  Q  D  K  E  R  Q  L  R  A  R  V  K  R  V  L  P  I  D  V  H
                            540                                                           600
CAGCCCCAGCCCCTGGGTGCTGGGAGCCCAGCTCCCCTGCCTGCTGACGCCCTGTCTGCCTTCTGCCTTGCTTGAGGCTGTGAGCCCAGATCTTGCCAGCTTTCAGCGGGCCCTGGACCAC
 Q  P  P  L  G  A  G  S  P  A  P  L  P  A  D  A  L  V  S  A  F  C  L  E  A  V  S  P  D  L  A  S  F  Q  R  A  L  D  H
                                    660                                                                            720
ATCACCACGCTGCTGAGGCCTGGGGGCCACTCCTCCTCATCGGGGCCCTGGAGGAGTCGTGGTACCTGGCTGGGTGCCAGTGTCTGACGGTGGTGCCAGTGTCTGAGGAGGAGGTGAGG
 I  T  T  L  L  R  P  G  G  H  L  L  L  I  G  A  L  E  E  S  W  Y  L  A  G  E  A  R  L  T  V  V  P  V  S  E  E  E  V  R
                                                   780                                                            840
GAGGCCCTGGTGCGTAGTGGCTACAAAGTCCGGGACCTCCGGACCTATATCATGCCTGCCCACCTTCAGACACAGGCGTAGATGATGTCAAGGGCGTCTTCTTCGCCTGGCTCAGAAGGTT
 E  A  L  V  R  S  G  Y  K  V  R  D  L  R  T  Y  I  M  P  A  H  L  Q  T  G  V  D  D  V  K  G  V  F  F  A  W  A  Q  K  V
                                                                                                                 900
GGGCTGTGAGGGGCTGTACCTGGTGCCCTGTGGCCCCCACCCACCTGGATTCCCTGTTCTTTGAAGTGGCACCTAATAAAGAAATAATACC(A)n
 G  L  ***
```

Fig. 7. Nucleotide sequence of cDNA encoding human PNMT type A (Kaneda et al., 1988; Sasaoka et al., 1989)

AMP responsive elements (GRE, CRE) near the transcripton initiation site of the DBH gene.

Human phenylethranslamine N-methyltransrase (PNMT) gene

PNMT is the terminal enzyme in catecholamine biosynthesis, and catalyzes the formation of adrenaline from noradrenaline, using S-adenosyl-L-methionine as the methyl donor (Axelrod, 1962). The enzyme is found in adrenaline cells of the adrenal medulla in which adrenaline is synthesized as a hormone, and in adrenaline neurons of the medulla oblongata and hypothalamus of the brain. The details of adrenaline neuron function in the brain are not yet clearly understood, but these adrenaline neurons may be involved in some important neurophysiological function such as cardiovascular and neuroendocrine regulations of the central nervous system.

We (Kaneda et al., 1988) reported the complete nucleotide sequence of a full-length human PNMT cDNA and the deduced amino acid sequence of the enzyme (Fig. 7). Determination of the nucleotide sequence revealed that human PNMT consists of 282-amino acid residues with a predicted molecular weight of 30,853 Da, including the intial methionine. The amino acid sequence of human PNMT was highly homologous (88%) with that of the bovine enzyme. We have also assigned the PNMT gene to chromosome 17 (Kaneda et al., 1988). Baetge et al. (1988) and we (Sasaoka et al., 1989) cloned the genomic DNA of human PNMT, and found that human PMNT gene to consist of three exons and two introns spanning about 2.2 kb.

We (Sasaoka et al., 1989) also observed the presence of a minor human PNMT mRNA (type B, 1.7 kb) besides the major mRNA (type A, 1.0 kb). Type B mRNA of human PNMT carries an approximately 700 nucleotide-long untranslated region in the 5'-terminus. This suggests that two types of mRNA are produced from a single gene through the use of two alternative promoters.

The 5'-flanking region of the human PNMT gene contains several consensus sequences for glucocorticoid responsive elements (GRE) and $Sp1$ binding sites.

Conclusion

We isolated and characterized cDNAs of all four human catecholamine-synthesizing enzymes, TH, AADC, DBH, and PNMT, and discovered multiple mRNAs of human TH, DBH and PNMT. The results are summarized in Table 1. The multiple mRNAs are produced by different alternative splicing mechanism from each single gene.

Four types of human TH are produced by two modes of alternative splicing; i.e., alternative use of two donor sites in a single exon, exon 1 (internal donor sites mechanism) and inclusion/exclusion of exon 2 (casette mechanism). They are different in the coding region due to the insertion/deletion sequence at the

T. Nagatsu et al.

Table 1. Characteristics of human catecholamine-synthesizing enzymes revealed by molecular cloning of cDNA and genomic DNA

Enzymes	cDNA	Base pairs	Modes of alternative mRNA splicing	Amino acid residues of subunit	Molecular weight of subunit	Subunit structure
Tyrosine Hydroxlase (TH)	type 1	1,491	internal donor sites and casette	497	55,533	Tetramer
	type 2	1,503		501	55,973	
	type 3	1,572		524	58,080	
	type 4	1,584		538	58,521	
Aromatic L-Amino Acid Decaarboxylase (AADC)		1,440		480	53,891	Dimer
Dopamine β-Hydroxylase (DBH)	type A	1,809	alternative polyadenylation siltes	578 (603)*	64,862	Tetramer
	type B	1,509				
Phenylethanolamine N-Methyltransferase (PNMT)	type A	846	altenative promoters	282	30,853	Monomer
	type R					

* The number in parenthesis includes the signal sequence

N-terminus. We have shown tissue-specific and high level expression of human TH gene to produce the multiple forms in the brain and adrenal gland. We found the multiplicity of TH in monkeys, suggesting that the multiplicity of TH mRNA is primate specific.

We found that in human DBH mRNA, the two types (types A and B) are produced by the alternative polyadenylation mechanism, and they are same in the coding region but different in the polyadenylation sites.

We also found two types of human PNMT mRNA (types A and B) which are produced by alternative promoter mechanism and have the same coding region, but different untranslated region in the 5' terminus.

Characterization of genes for human catecholamine-synthesizing enzymes may clarify the molecular mechanisms underlying the differentiation and expression of catecholaminergic neurons during development and aging, and the regulation of catecholamine neurotransmitter biosynthesis, both of which are involved in physiological and pathological conditions including Parkinson's disease.

Acknowledgment

We are grateful to the support by a Grant-in-Aid for Scientific Research on Priority Areas, "Molecular Biology of the Motor System", Ministry of Education, Science and Culture, Japan.

References

Axelrod J (1962) Purification and properties of phenylethanolamine N-methyltransferase. J Biol Chem 237:1657–1660

Baetge EE, Behringer RR, Messsing A, Brinster RL, Palmiter RD (1988) Transgenic mice express the human phenylethanolamine N-methyltransferase gene in adrenal medulla and retina. Proc Natl Acad Sci USA 85:3648–3652

Brenneman AR, Kaufman S (1964) The role of tetrahydropteridines in the enzymatic conversion of tyrosine to 3,4-dihydroxyphenylalanine. Biochem Biophys Res Commun 17:177–183

Coker GT III, Vinnedge L, O'Malley KL (1988) Characterization of rat and human tyrosine hydroxylase genes: functional expression of both promoters in neuronal and non-neuronal cell types. Biochem Biophys Res Commun 157:1341–1347

Dahlstrom A, Belmaker RH, Sandler M (eds) (1988) In: Progress in catecholamine research, part A. Basic aspects and peripheral mechanisms. Alan R Liss, New York, pp 1–613

Friedman S, Kaufman S (1965) 3,4-Dihydroxyphenylethylamine β-hydroxylase. Physical properties, copper content, and role of copper in the catalytic activity. J Biol Chem 240:4763–4773

Fujisawa H, Okuno S (1987) Tyrosine 3-monooxygenase from rat adrenals. In: Kaufman S (ed) Methods in enzymology, vol 142. Academic Press, New York, pp 63–71

Ginns EI, Rehari M, Martin BM, Weller M, O'Malley KL, La Marca ME, McAllister CG, Paul SM (1988) Expression of human tyrosine hydroxylase cDNA in invertebrate cells using a baculovirus vector. J Biol Chem 263:7406–7410

Grima B, Lamouroux A, Blanot F, Faucon Biguet N, Mallet J (1985) Complete mRNA coding sequence of rat tyrosine hydroxylase. Proc Natl Acad Sci USA 82:617–621

Grima B, Lamouroux A, Boni C, Julien JF, Javoy-Agid F, Mallet J (1987) A single human gene encoding multiple tyrosine hydroxylases with different predicted functional characteristics. Nature 326:707–711

Horellou P, Le Bourdelles B, Clot-Humbert J, Guibert B, Leviel V, Mallet J (1988) Multiple human tyrosine hydroxylase enzymes, generated through alternative splicing, have different specific activities in Xenopus oocytes. J Neurochem 51:652–655

Ichikawa S, Ichinose H, Nagatsu T (1990) Multiple mRNAs of monkey tyrosine hydroxylase. Biochem Biophys Res Commun 173:1331–1336

Ichinose H, Kurosawa Y, Titani K, Fujita K, Nagatsu T (1989) Isolation and characterization of a cDNA clone encoding human aromatic L-amino acid decarboxylase. Biochem Biophys Res Commun 164:1024–1030

Ishii A, Hagihara M, Matsuura S, Uchida K, Kiuchi K, Kaneda N, Toya S, Kohsaka S, Nagatsu T (1990) Effect of (6R)- and (6S)-tetrahydrobiopterin on L-3,4-dihydroxyphenylalanine (DOPA) formation in NRK fibroblasts transfected with human tyrosine hydroxylase type 2 cDNA. Neurochem Int 17:625–632

Kaneda N, Kobayashi K, Ichinose H, Kishi F, Nakazawa A, Kurosawa Y, Fujita K, Nagatsu T (1987) Isolation of a novel cDNA clone for human tyrosine hydroxylase: alternative RNA splicing produces four kinds of mRNA from a single gene. Biochem Biophys Res Commun 146:971–975

Kaneda N, Ichinose H, Kobayashi K, Oka K, Kishi F, Nakazawa A, Kurosawa Y, Fujita K, Nagatsu T (1988) Molecular cloning of cDNA and chromosomal assignment of the gene for human phenylethanolamine N-methyltransferase, the enzyme for epinephrine biosynthesis. J Biol Chem 263:7672–7677

Kaneda N, Sasaoka T, Kobayashi K, Katsuki M, Yokoyama M, Nagatsu I, Kurosawa Y, Fujita K, Nagatsu T (1990) Production and analysis of transgenic mice carrying human tyrosine hydroxylase gene (in Japanese). Seikagaku (Jap Biochem Soc) 62:975–975

Kaneda N, Kobayashi K, Ichinose H, Sasaoka T, Ishii A, Kiuchi K, Kurosawa Y, Fujita K, Nagatsu T (1990) Molecular biological approaches to catecholamine neurotransmitters and brain aging. In: Nagatsu T, Hayaishi O (eds) Aging of the brain. Cellular and molecular aspects of brain aging and Alzheimer's disease. Japan Scientific Press, Tokyo, and Karger, Basel, pp 53–56

Kaneda N, Sasaoka T, Kobayashi K, Kiuchi K, Nagatsu I, Kurosawa Y, Fujita K, Yokoyama M, Nomura T, Katsuki M, Nagatsu T (1991) Tissue-specific and high-level expression of human tyrosine hydroxylase gene in transgenic mice. Neuron 6:1–12

Kobayashi K, Kaneda N, Ichinose H, Kishi F, Nakazawa A, Kurosawa Y, Fujita K, Nagatsu T (1987) Isolation of a full length cDNA clone encoding human tyrosine hydroxylase type 3. Nucl Acids Res 15:6733–6733

Kobayashi K, Kaneda N, Ichinose H, Kishi F, Nakazawa A, Kurosawa Y, Fujita K, Nagatsu T (1988a) Structure of the human tyrosine hydroxylase gene: alternative splicing from a single gene accounts for generation of four mRNA types. J Biochem 103:907–912

Kobayashi K, Kiuchi K, Ishii A, Kaneda N, Kurosawa Y, Fujita K, Nagatsu T (1988b) Expression of four types of human tyrosine hydroxylase in COS cells. FEBS Lett 238:431–434

Kobayashi K, Kurosawa Y, Fujita K, Nagatsu T (1989) Human dopamine β-hydroxylase gene: two mRNA types having different 3'-terminal regions are produced through alternative polyadenylation. Nucl Acids Res 17:1089–1102

Kojima K, Mogi M, Oka K, Nagatsu T (1984) Purification and immunochemical characterization of human adrenal tyrosine hydroxylase. Neurochem Int 6:475–480

Lamouroux A, Vigny N, Facon Bigunet MC, Darmon R, Franck R, Henry J-P, Mallet J (1987) The primary structure of human dopamine β-hydroxylase: insights into the relationship between the soluble and the membrane-bound forms of the enzyme. EMBO J 6:3931–3937

Le Bourdellès B, Boularand S, Boni C, Horellou P, Dumas S, Grima B, Mallet J (1988) Analyses of the 5'-region of the human tyrosine hydroxylase gene: combinatorial patterns of exon splicing generate multiple regulated tyrosine hydroxylase isoforms. J Neurochem 50:988–991

Lewis EJ, Harrington CA, Chikaraishi DM (1987) Transcriptional regulation of the tyrosine hydroxylase gene by glucocorticoid and cyclic AMP. Proc Natl Acad Sci USA 84:3550–3554

Lovenberg W, Weissbach H, Udenfriend S (1962) Aromatic L-amino acid decarboxylase. J Biol Chem 237:89–93

Man in't Veld AJ, Boomsma F, Moleman P, Schalekamp MADH (1987) Congenital dopamine-beta-hydroxylase deficiency. A novel orthostatic syndrome. Lancet i: 183–187

Matuura S, Sugimoto T, Murata S, Sugawara Y, Iwasaki H (1985) Stereochemistry of biopterin cofactor and facile methods for the determination of the stereochemistry of a biologically activate 5,6,7,8-tetrahydropterin. J Biochem 98:1341–1348

Mogi M, Harada M, Kiuchi K, Kojima K, Kondo T, Narabayashi H, Rausch D, Riederer P, Jellinger K, Nagatsu T (1988) Homospecific activity (activity per enzyme protein) of tyosine hydroxylase increases in parkinsonian brain. J Neural Transm 72:77–81

Mogi M, Kojima K, Nagatsu T (1984) Detection of inactive or less active forms of tyrosine hydroxylase in human brain and adrenals by a sandwich enzyme immunoassay. Anal Biochem 138:125–132

Mogi M, Kojima K, Harada M, Nagatsu T (1986) Purification and immunological properties of tyrosine hydroxylase in human brain. Neurochem Int 8:423–428

Nagatsu T (1977) Dopamine-β-hydroxylase in blood and cerebrospinal fluid. Trends Biochem Sci 2:217–219

Nagatsu T, Kato T, Numata (Sudo) Y, Ikuta K, Sano M, Nagatsu I, Kondo Y, Inagaki S, Iizuka R, Hori A, Narabayashi H (1977) Phenylethanolamine N-methyltransferase and other enzymes of catecholamine metabolism in human brain. Clin Chim Acta 75:221–232

Nagatsu T, Levitt M, Udenfriend S (1964) Tyrosine hydroxylase. The initial step in norepinephrine biosynthesis. J Biol Chem 239:2910–2917

Nagatsu T, Oka K (1987) Tyrosine 3-monooxygenase from bovine adrenal medulla. In: Kaufman S (ed) Methods in enzymology, vol 142. Academic Press, New York, pp 56–62

Nagatsu T, Yamaguchi T, Kato T, Sugimoto T, Matsuura S, Akino M, Nagatsu I, Iizuka R, Narabayashi H (1981) Biopterin in human brain and unine from controls and parkinsonian patients: application of a new radioimmunoassay. Clin Chim Acta 109:305–311

Nagatsu T, Yamaguchi T, Rahman MK, Trocewicz J, Oka K, Hirata Y, Nagatsu I, Narabayashi H, Kondo K, Iizuka R (1984) Catecholamine-related enzymes and the biopterin cofactor in Parkinson's disease and related extrapyramidal diseases. In: Hassler RG, Christ JF (eds) Advances in neurology, vol 40. Raven Press, New York, pp 467–473

O'Malley KL, Anhalt MJ, Martin BM, Kelsoe JR, Winfield SL, Ginns EI (1987) Isolation and characterization of the human tyrosine hydroxylase gene: identification of 5'-alternative splice sites responsible for multiple mRNAs. Biochemistry 26:6910–6914

Robertson D, Goldberg MR, Onrot J, Hollister AS, Wiley R, Thompson JR, Robertson RM (1986) Isolated failure of autonomic noradrenergic neurotransmission. N Engl J Med 314:1494–1497

Sasaoka T, Kaneda N, Kurosawa Y, Fujita K, Nagatsu T (1989) Structures of human phenylethanolamine N-methyltransferase gene: existence of two types of mRNA with different transcription initiation sites. Neurochem Int 15:555–565

Sumi C, Ichinose H, Nagatsu T (1990) Characterization of recombinant human aromatic L-amino acid decarloxylase expressed in COS cells. J Neurochem 55:1075–1078

Tank AW, Weiner N (1987) Tyrosine 3-monooxygenase from rat pheochromocytoma. In: Kaufman S (ed) Methods in enzymology, vol 142. Academic Press, New York, pp 71–82

Uchida K, Takamatsu K, Kaneda N, Toya S, Tsukada Y, Kurosawa Y, Fujita K, Nagatsu T, Kohsaka S (1988) Transfection of tyrosine hydroxylase cDNA into C6 cell. Proc Jpn Acad Ser B 64:290–293

Uchida K, Takamatsu T, Kaneda N, Toya S, Tsukada Y, Kurosawa Y, Fujita K, Nagatsu T, Kohsaka S (1989) Synthesis of L-3,4-dihydroxyphenylanin by tyrosine hydroxylase cDNA-transfected C6 cells: application for intracerebral grafting. J Neurochem: 53:728–732

Uchida K, Ishii A, Kaneda N, Toya S, Nagatsu T, Kohsaka S (1990a) Tetrahydrobioptenin-dependent production of L-DOPA in NRK fibroblasts fransfected with tyrosine hydroxylase cDNA: future use for intracerebral grafting. Neurosci Lett 109:282–286

Uchida K, Toya S, Tsukada S, Nagatsu T, Kohsaka S (1990b) Transfection of tyrosine hydroxylase cDNA into non-neuronal cells: application for intracerebral grafting. In: Nagatsu T, Hayaishi O (eds) Aging of the brain. Cellular and molecular aspects of brain aging and Alzheimer's disease. Japan Scientific Societies Press, Tokyo, and Karger, Basel, pp 79–93

Zigmond RE, Schwarzchild MA, Rittenhouse AR (1989) Acute regulation of tyrosine hydroxylase by nerve activity and by neurotransmitters via phosphorylation. Ann Rev Neurosci 12:415–416

Correspondence: Prof. Dr. T. Nagatsu, Department of Biochemistry, Nagoya University School of Medicine, 65 Tsurumai-cho, Showa-ku, Nagoya 466, Japan.

Receptor alterations in Parkinson's disease

N. Ogawa and M. Asanuma

Department of Neurochemistry, Institute for Neurobiology, Okayama University Medical School,
Okayama, Japan

Summary

Not only dopamine but also enzymes involved in catecholamine synthesis and other biogenic amines have been found to decrease in the brain of PD patients; amino acid and neuropeptide levels also appear to be affected. Furthermore, changes in various neurotransmitter receptors have been detected by receptor studies. These studies are now effectively employed for the exploration of administration modalities with minimum side effects and for the development of new drugs as well as for the investigation of the action mechanisms of available agents.

Introduction

Although the primary biochemical change in Parkinson's disease (PD) is a dopamine (DA) deficiency, recent studies have suggested that other neurotransmitters as well as their receptors show changes in PD patients. Neurotransmitter concentrations alone do not always accurately reflect the pathophysiology in the patients. The measurable level of a neurotransmitter is the sum of the substance being synthesized, stored, and released, and a large storage capacity may conceal fluctuations of minute released amounts. Synaptic receptors, on the other hand, are localized at the synaptic junctions and are considered to change in closer association with the condition of the synapse, showing supersensitivity in response to a reduction of the neurotransmitter release and subsensitivity to its increase. Although determinations of both neurotransmitter and receptor levels are indispensable for the evaluation of synaptic activities, the later levels are often more sensitive.

In this paper, changes in neurotransmitters and their receptors in parkinsonian patients are summarized, and the therapeutic application of receptor study techniques is discussed.

Transmitters and drug-induced parkinsonism

Before discussing the receptors, the current status of the relevant neuro-transmitters will be summarized.

The primary biochemical change in PD has been identified to be DA deficiency in nigro-striatal dopaminergic neurons, but several other biochemical changes are now known to occur.

The decrease in DA, noradrenaline (NE), serotonin (Ehringer and Horny-kiewicz, 1960), and homovanillic acid (HVA) (Bernheimer et al., 1973) levels is marked. Enzymes such as tyrosine hydroxylase (TH), dopa-decarboxylase,

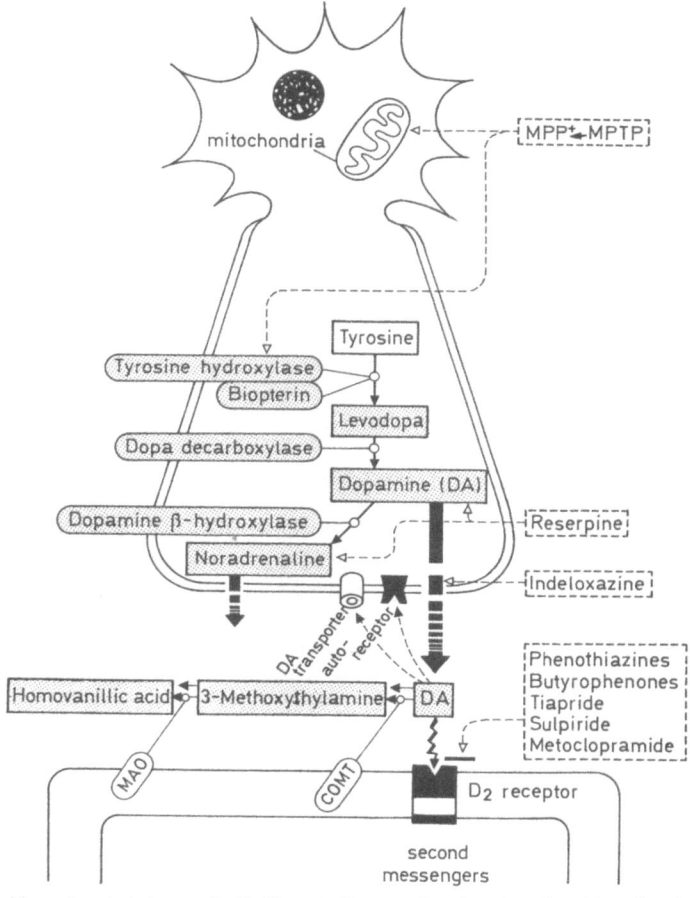

Fig. 1. Neurochemical changes in Parkinson's disease and action sites of parkinsonism inducing drugs. Shadowed amines or enzymes are reduced in the brain of Parkinson's disease. ▭ : amines and metabolites; ⬭ : enzymes; ⸢⸉ : drugs which induce parkinsonism; ⟶ : change; ⟶○ : action sites of enzymes; ⟶▷ : inhibit

dopamine-β-hydroxylase, and phenylethanolamine N-methyltransferase, especially rate-limiting TH (McGeer and McGeer, 1976; Nagatsu et al., 1977), are involved in catecholamine synthesis. Biopterine, a coenzyme of TH, is also decreased in the striatum of PD patients (Nagatsu et al., 1981). On the other hand, the brain concentrations of monoamine oxidase and catechol-o-methyltransferase, which are catecholamine-metabolizing enzymes, have been reported to show no differences between patients and controls (Nagatsu et al., 1977). The above findings indicate that the reduction in catecholamine levels in the brain of PD patients is due to the decreased biosynthesis of catecholamines (Fig. 1).

Choline acetyltransferase, a marker for acetylcholine, is reported to be normal or decreased (McGeer and McGeer, 1976; Reisine et al., 1977) while the serotonin concentration is decreased (Bernheimer et al., 1961). GABA was considered to decrease, but Perry et al. (1983) noted an increase in the PD brain. As for neuropeptides in the brain of PD patients, substance P, somatostatin, methionine-enkephalin, and cholecystokinin-octapeptide levels are lower (Studler et al., 1982; Taquet et al., 1982; Mauborgne et al., 1983), but arginine-vasopressin and TRH levels are comparable to those in the control brain.

The pathophysiological significance of these neurotransmitter changes is obscure. The DA system has been analyzed more extensively than any other neurotransmitter systems because of the critical depletion of DA in PD and the therapeutic response to dopaminomimetic therapy. Drugs which decrease the DA levels, such as 1-methyl-4-phenyl-1,2,3,6-tetrahydropyridine (MPTP), reserpine or indeloxazine, have been shown to produce drug-induced parkinsonism (Langston et al., 1983; Ogawa, 1989) (Fig. 1).

Receptors

The receptor studies can be of value in identifying (a) receptor changes in PD brain, (b) the primary site of action of antiparkinsonian drugs, (c) secondary sites that may produce side effects, (d) long-term receptor changes induced by drugs, and (e) drugs that enhance or normalize pathological levels of receptors.

DA receptors

DA receptors, which are considered to have the greatest clinical implications in PD, are classified into two subtypes (Creese and Left, 1982): D1 receptors linked to DA-sensitive adenylate cyclase are responsive to micromolar concentrations of both DA agonists and antagonists; D2 receptors (DA/neuroleptic receptors) are sensitive to nanomolar concentrations of DA antagonists but to variable levels of DA agonists. Other subtype autoreceptors (DA agonist sites) are located in the DAnergic nerve terminals and are sensitive to nanomolar concentrations of DA agonists and to micromolar levels of DA antagonist.

Table 1. Receptor alterations in discrete brain regions in Parkinson's disease

	SN	CN	PUT	PAL	ACC	FCX	LIMCX	Reference
DOPAMINE D1 R								
[3H]-flupenthixol								
PD total	=	=	←	=	=			Rinne JO 1985
PD levodopa-	→	=	=	=	=			
PD levodopa+	=	=	←	←	=			
[3H]-SCH23390								
PD	SNC →	=	=					Pimoule C 1985
PD	SNR =	=	=					Cash R 1987
DOPAMINE D2 R								
[3H]-spiroperidol								
PD	=	→	=					Reisine TD 1977
PD	=	→	=					Winkler MH 1980
PD	→	→←←→	→←←→	→←←→	←			Rinne UK 1980,1981
PD levodopa- p.m.53h / p.m.14h neuroleptics+		=←	=	= ←			= =	
PD levodopa+ 1260mg/day 3.6y / 865mg/day 2.1y neuroleptics+		= ←	=	= ←	= ←		= =	
PD levodopa- / PD levodopa+	=	= / =						Spokes EGS 1981
PD levodopa 2-8y	=	=	←		=			Bokobza B 1984
PD			←	→	←	=		Javoy-Agid F 1984
PD		=	←		=			Pimoule C 1985
PD levodopa- / PD levodopa+		← =	← =					Guttman M 1985
PD levodopa- / PD levodopa+		=	=					Guttman M 1986
PD		=	=					Cash R 1987
[3H]-haloperidol								
PD levodopa-		=	←					Lee T 1978
PD levodopa+		=	=					
DOPAMINE AUTORECEPTOR		CN	PUT					
[3H]-apomorphine								
PD		=	→					Lee T 1987

Measure	CN	PUT	STR	FCX	LC	Reference
PD levodopa-NL-	→	=				Rinne UK 1983
levodopa+NL-	→	→				
levodopa-NL+	=	=				
levodopa+NL+	=	=				
DOPAMINE TRANSPORTER						
[3H]-cocaine PD	→	→				Pimoule C 1985
[3H]-GBR12935 PD	→					Janowsky A 1987
PD		→				Hirai M 1988
PD			↓→			Maloteaux JM 1988
PD			→			Pearce RKB 1990
ALPHA-1 ADRENERGIC R						
[3H]-prazosin						Cash R 1984
PD total				=		
PD non-demented				=		
PD demented				↑		
ALPHA-2 ADRENERGIC R						
[3H]-clonidine						Cash R 1984
PD total				→		
PD non-demented				=		
PD demented				←		
[3H]-rauwolscine						Cash R 1987
PD non-demented					=	
PD demented					=	
BETA-1 ADRENERGIC R						
[3H]-DHA						Cash R 1984
PD total				←		
PD non-demented				=		
PD demented				=		
BETA-2 ADRENERGIC R						
[3H]-DHA						Cash R 1984
PD total				=		
PD non-demented				=		
PD demented				=		

	SN	SNR	CN	PUT	PAL	ACC	FCX	PARCX	Reference
5-HT 1 R									
[3H]-serotonin									
PD	=								Reisine TD 1977
PD total			=	→	=		=		Rinne UK 1980
PD levodopa-			=	=	=		=		
PD levodopa+			=	=	=				
PD						=			Kienzl B 1981
PD		=	=	=	=	=	→ =		Waeber C 1989
5-HT 2 R									
[3H]-ketanserin									
PD non-demented								= →	Perry EK 1984
PD demented									
ACETYLCHOLINE M1+M2 R									
[3H]-QNB									
PD	=		=	←	=	=	=		Reisine TD 1977
PD levodopa- low D2 binding			→	→			←		Rinne UK 1980
high D2 binding			←	←	←		←		
PD levodopa+ low D2 binding			=	=	=		←		
high D2 binding			=	=					
PD	=		=			=	←		Ruberg M 1982
PD non-demented AC-			=				=		Dubois B 1983
PD non-demented AC+			←				←		
PD demented AC-			←				←		
PD demented AC+							←		
PD	=		=	=	=		=		Penney JB 1987
PD total	=		=	=	=		=		Nishino N 1988
PD demented	←		←	←	←		←		
ACETYLCHOLINE M1 R									
[3H]-pirenzepine									
PD				=					Nishino N 1988
ACETYLCHOLINE NICOTINIC R									
[3H]-acetylcholine									
PD							→		Whitehouse PJ 1988

	SN	SNC	SNR	CN	PUT	PAL	FCX	HIPPO	Reference
GABA R [3H]-GABA									
PD	↓						↑	→	Lloyd KG 1977
PD				=	=	=			Reisine TD 1977
PD total	→			=	=	=	=		Rinne UK 1980
PD levodopa-	=			=	=	=	=		
PD levodopa+	→			=	=	=	=		
GABA-A R [3H]-muscimol									
PD	=					=	→		Penney JB 1987
PD	→			→	→	→			Nishino N 1988
BDZ R [3H]-flunitrazepam									
PD	↓			=	=	=			Penney JB 1987
PD total	=			←			=		Maloteaux JM 1988
PD non-demented							=		
PD moderate-demented							=		
PD severe-demented							=		
BDZ R type I [3H]-flunitrazepam									
PD	→								Uhl GR 1986
BDZ R type II [3H]-flunitrazepam									
PD	=								Uhl GR 1986
GLYCINE R [3H]-strychnine									
PD		↓	→	→					de Montis G 1982
OPIATE mu R [3H]-naloxone									
PD	↓			→					Reisine TD 1979
PD levodopa-	=			=	=			=	Rinne UK 1983
PD levodopa+	=				=			=	

Ligand / Receptor	Condition	SN / SNC	CN	PUT	PAL	ACC	FCX	LIMCX	HIPPO	Reference
[3H]-dihydromorphine	PD	↓								Uhl GR 1986
[3H]-DAGO	PD	=	=	=	=					Delay-Goyet P 1987
OPIATE kappa R [3H]-ethylketocyclazocine	PD	↓ (SN)	=	=			= (FCX)			Uhl GR 1986
[3H]-etrophine	PD	=	=	=	=		=			Delay-Goyet P 1987
OPIATE delta R [3H]-Leu Enk	PD total	(SNC)	=	↑		=	=	↑	↑	Rinne UK 1980
	PD levodopa–		=	↑		=	=	↑	=	
	PD levodopa+		=	↑		=	=			
[3H]-Met Enk	PD	↑	↑	↑		↑			↑	Rinne UK 1983
[3H]-D-Ala,Met Enk	PD	↓	=							Llorens-Cortes C 198
SOMATOSTATIN R [125I]-Leu8dTry-TyrSS28	PD	↓ (SN)								Uhl GR 1986
NEUROTENSIN R [125I]-Tyr NT	PD	↓ (SN)								Sadoul JL 1984
[3H]-NT	PD	↓								Uhl GR 1986

AC anticholinergic therapy; *ACC* accumbens nucleus; *CN* caudate nucleus; *FCX* frontal cortex; *HIPPO* hippocampus; *LC* locus ceruleus; *LIMCX* limbic cortex; *NL* neuroleptics; *PAL* globus pallidus; *PARCX* parietal cortex; *p.m.* postmortal intervals; *PUT* putamen; *SN* substantia nigra; *SNC* substantia nigra pars compacta; *SNR* substantia nigra pars reticulata

↑ increased; ↓ decreased; = not affected

Both D1 and D2 receptors are present in the striatum, and the accummulating evidence indicates that blockade of the D2 receptor produces parkinsonism and that the stimulation of the D2 receptor is of prime importance in the effectiveness of antiparkinsonian therapy.

Different ligands have been employed to study DA receptors of PD brains. These have included [^{3}H]spiroperidol, [^{3}H]haloperidol and [^{3}H]apomorphine.

Table 1 summarizes the changes in striatal DA receptor levels in parkinsonism. Lee et al. (1978) studied the D2 receptor concentration in the striatum of the brain of parkinsonian patients using [^{3}H]haloperidol as a marker for D2 receptor binding, and suggested an increase in D2 receptors in patients not administered levodopa but no changes in those treated with the agent until death. Rinne et al. (1978, 1981, 1982) studied D_2 binding with [^{3}H]spiroperidol, and reported increases in D_2 receptors in some patients treated with levodopa but decreases in others. However, the results of such assays may be influenced by the severity or duration of the disease, and low binding may correlate with advanced disease and loss of response to lovodopa. Guttman and Seeman (1985), also using [^{3}H]spiperone, observed increased D2 receptor binding in the striatum of PD patients who had not received levodopa and normalization of D2 receptor binding shortly after levodopa treatment. Guttman et al. (1986) reported that the D2 receptor densities in the striatum from 36 PD patients did not change with the patients' age, duration of disease or duration of the levodopa treatment, and suggested that the depletion of D2 receptors was not the cause of the clinical deterioration evident after long-term levodopa treatment. Similarly, there are various reports on D2 receptor in PD patients. In contrast, animal studies demonstrated increased receptor numbers following lesions of the nigrostriatal tract (Lang and Blair, 1984).

No agreement has been reached on the behavior of DA sensitive adenylate cyclase (D1 receptors), with some investigators reporting increases (Nagatsu et al., 1978) and others decreases (Shibuya, 1979). Direct D1 receptor binding assays using the D1 receptor ligand [^{3}H]SCH 23390 has disclosed that the D1 receptor density was unchanged in the putamen and the substantia nigra pars reticulata of PD patients but was decreased by 28% in the substantia nigra pars compacta of PD patients (Cash et al., 1987). Autoreceptors were found to be reduced in the brain of PD patients regardless of the administration of levodopa (Lee et al., 1978).

Other monoaminergic receptors were less affected in PD. The level of α_2 adrenergic receptors was slightly low and that of adrenergic receptors significantly high (+28%) in the frontal cortex of parkinsonian subjects. The level of α_1 adrenergic receptors in demented parkinsonians was slightly higher than that in nondemented patients (Cash et al., 1984). Serotonin (5-HT$_1$) receptors detected using [^{3}H]-serotonin showed no changes.

Muscarinic cholinergic receptors (MCR)

In PD, the striatal cholinergic system was strongly dependent on the function of the dopaminergic system. Findings on striatal MCR are conflicting.

Rinne et al. (1980, 1981) noted decreased MCR ([^{3}H]QNB binding) with low [^{3}H]spiroperidol binding, while increased MCR has been reported by Reisine et al. (1977) and Nishino et al. (1988) reported no change in MCR.

Amino acid receptors

GABA receptor binding showed no changes in the striatum (Rinne et al., 1980) but a marked decrease in the striatum, substantia nigra and frontal cortex (Nishino et al., 1988).

The amount of glycine receptors detected by [^{3}H]strychinine was decreased in the substantia nigra (de Montis et al., 1982).

Opioid receptors

There was a significant depletion of μ-receptors in the caudate nucleus ([^{3}H]naloxone binding) (Reisine et al., 1979), while δ-receptors ([^{3}H]leu- and [^{3}H]met-enkephalin binding) were increased (Rinne et al., 1980), corresponding to the depletion of enkephalins (Taquet et al., 1983).

Peptide receptors

There are reports of strong reduction of [^{3}H]neurotensin binding in the striatum, substantia nigra and cerebral cortex (Quirion et al., 1982; Sadoul et al., 1984; Uhl et al., 1984). Somatostatin receptors and neurotensin receptors showed low levels in the substantia nigra (Uhl et al., 1986).

Therapeutic application of receptor studies

Many CNS-acting drugs, including antiparkinsonian agents, act on several receptors. Interestingly, levodopa showed crossreactivity in a [^{3}H]DA-radio-labeled receptor assay (RRA) system (Ogawa et al., 1984). Moreover, the displacement curves of levodopa against [^{3}H]DA-receptor binding paralleled that of DA, suggesting that levodopa may competitively bind to DA binding sites. In addition, from the absence of transformation of levodopa to DA confirmed by HPLC following incubation of levodopa with the synaptic membrane under the same condition as the RRA system, levodopa was found to act on DA receptors in an unmodified form. This suggests that at least some of levodopa-induced dyskinesia develops due to the action of levodopa itself on DA receptors. This is in line with the increased incidence of the dyskinesia following the introduction of a combination of a peripheral dopa-decarboxylase inhibitor with levodopa therapy as well as with the speculation by Klawans et al. (1975, 1977) that the dyskinesia is caused by levodopa itself.

Bromocriptine has recently been used as a therapeutic agent for PD, but the mechanism of its antiparkinsonian effect has not fully been clarified. As previously reported (Ogawa et al., 1981), the agent has far greater affinity to enkephalin binding sites than to DA binding sites. Since enkephalin neurons are known to terminate in the nerve endings of DA neurons in the striatum, the action site of bromocriptine is considered to be different from that of DA. This suggests that at least part of the therapeutic antiparkinsonian effect of bromocriptine is due to its allosteric effects on the enkephalin receptor, which in turn influences the function of striatal DA neurons and/or striatal cholinergic neurons. Furthermore, the effects of chronic levodopa and bromocriptine treatment on striatal DA receptor were investigated in rats with parkinsonism induced experimentally by a unilateral intrastriatal microinjection of 6-hydroxydopamine (6-OHDA) (Ogawa et al., 1984). After 3 weeks, the amount of specific binding of [^{3}H]DA increased by 200% on the treated side compared with the untreated side. Scatchard analysis revealed that the 6-OHDA lesion increased the maximal number of DA binding sites, which indicates that at least part of the denervation supersensitivity mechanism is due to an increase in the number of striatal DA receptors. Repeated administration of levodopa for 3 weeks resulted in a great reduction of striatal DA receptors on the treated side to almost the same level as on the untreated side of the control group. On the other hand, 3 weeks of administration of bromocriptine did not reduce the DA receptors. Interestingly, treatment with levodopa and bromocriptine in combination had no significant effect on the specific binding of [^{3}H]DA compared with the treated side of the control group (Ogawa et al., 1984). This suggests that bromocriptine is effective in blocking the downregulation of DA receptors evoked by levodopa. It may be possible to obtain better therapeutic results in parkinsonism by combination therapy of bromocriptine and levodopa because of their different action mechanisms in the striatum. Furthermore, the addition of bromocriptine to the therapeutic regimen of levodopa-treated patients may overcome the shortcomings, including the "wearing-off" phenomenon, which is seen late in chronic levodopa therapy. Favorable long-term control of parkinsonian symptoms may also be achieved by a combination of levodopa with low doses of bromocriptine (Teychenne et al., 1981).

References

Bernheimer H, Birkmayer H, Hornykiewicz O (1961) Verteilung des 5-Hydroxytryptamins (Serotonin) im Gehirn des Menschen und sein Verhalten bei Patienten mit Parkinson-Syndrom. Klin Wochenschr 39:1056–1059

Bernheimer H, Birkmayer W, Hornykiewicz O, Jellinger K, Seitelberger F (1973) Brain dopamine and the syndrome of Parkinson and Huntington. Clinical, morphological and neurochemical correlations. J Neurol Sci 20:425–455

Creeogse I, Lelf SE (1982) Dopamine receptors: a classification. J Clin Psychopharmacol 2:329–335

de Montis G, dBeaumont K, Jovoy-Agid F, Agid Y, Constandinidis J, Lowenthal A, Lloyd KG (1982) Glycine receptors in the human substantia nigra as defined by [^{3}H]strychinine binding. J Neurochem 38:718–724

Ehringer H, Hornykiewicz O (1960) Verteilung von Noradrenalin und Dopamin (3-Hydroxytryramin) in Gehirn des Menschen und ihr Verhalten bei Erkrankungen des extrapyramidalen Systems. Klin Wochenschr 38:1236–1239

Guttman M, Seeman P (1987) Dopamine D2 receptor density in parkinsonian brain is constant for duration of disease, age, and duration of L-DOPA therapy. In: Yahr MD, Bergmann (eds) Advance in neurology, vol 45. Raven Press, New York, pp 51–57

Klawans HL, Crosset P, Dana N (1975) Effect of chronic amphetamine exposure on stereotyped behavior: implications for pathogenesis of L-dopa-induced dyskinesia. In: Calne DB, Chase TN, Barbea A (eds) Advances in neurology, vol 9. Raven Press, New York, pp 105–112

Klawans HL, Goetz C, Nausieda PA, Weiner WJ (1977) Levodopa-induced dopamine receptor hypersensitivity. Ann Neurol 2:125–129

Langston, JW, Ballard P, Tetrud JW, Irwin I (1983) Chronic parkinsonism in humans due to a product of meperidine-analog synthesis. Science 219:979–980

Lee T, Seeman P, Rajput A, Farley IJ, Hornykiewicz O (1978) Receptor basis for dopaminergic supersensitivity in Parkinson's disease. Nature 273:59–61

Long A, Blair R (1984) Parkinson's disease in 1984: an update. Can Med Assoc J 131:1031–1037

McGeer PL, McGeer EG (1976) Enzymes associated with the metabolism of catecholamines, acetylcholine and GABA in human controls and patients with Parkinson's disease and Huntington's chorea. J Neurochem 26:65–76

Mauborgne A, Javoy-Agid F, Legrand JC, Agid Y, Cesselin F (1983) Decrease of substance P-like immunoreactivity in the substantia nigra and pallidum of parkinsonian brain. Brain Res 268:167–170

Nagatsu T, Kanamori T, Kato T, Iizuka R, Narabayashi H (1978) Dopamine-stimulated adenylate cyclase activity in the human brain: changes in parkinsonism. Biochem Med 19:360–365

Nagatsu T, Kato T, Nagatsu(sudo) Y, Ikuta K, Sano M, Nagatu I, Kondo Y, Inagaki S, Iizuka R, Hori A, Narabayashi H (1977) Phenylethanolamine N-methyltransferase and other ezymes of catecholamine metabolism in human brain. Clin Chim Acta 75:221–232

Nagatsu T, Yamaguchi T, Kato T, Sugimoto T, Matsuura S, Akino M, Nagatsu I, Iizuka R, Narabayashi H (1981) Biopterine in human brain and urine from controls and parkinsonian patients: application of a new radioimmunoassay. Clin Chim Acta 109:305–311

Ogawa N (1989) Parkinsonism induced by indeloxazine hydrochloride in the elderly. Clin Ther 11:802–806

Ogawa N, Mizuno S, Mori A, Kuroda H (1984) Chronic dihydroergotoxine administration sets on receptors for enkephalin and thyrotropin releasing hormone in the aged-rat brain. Peptides 5:53–56

Ogawa N, Yamawaki Y, Kuroda H, Ofuji T (1981) Effects of bromocriptine on receptor binding of methionine-enkephalin. Neurosci Lett 23:215–218

Ogawa N, Yamawaki Y, Kuroda H, Takayama H, Ota Z (1984) Differences in the effect of levodopa and bromocriptine on rat striatal dopamine receptors. Neurosciences (Kobe) 10:259–266

Perry TL, Javoy-Agid F, Agid Y, Fibiger HC (1983) Striatal GABAergic neuronal activity is not reduced in Parkinson's disease. J Neurochem 40:1120–1123

Quirion R, Larson TA, Calne D, Chase T, Rioux F, St-Pierre S, Evarist H, Pert A, Pert CB (1982) Analysis of [^{3}H]neurotensin receptors by computed densitometry: visualization of control and peripheral neurotensin receptors. In: Nemeroff CB, Prange

AJ (eds) Neurotensin: a brain and gastrointestinal peptide. Ann NY Acad Sci 400:415–417

Reisine TP, Fields JZ, Yamamura HI, Bird ED, Spokes E, Schreiner PS, Enna SJ (1977) Neurotransmitter receptor alterations in Parkinson's disease. Life Sci 21:335–344

Reisine TD, Rosser M, Spokes E, Iversen L, Yamamura HI (1979) Alterations in brain opiate receptors in Parkinson's disease. Brain Res 173:378–392

Rinne UK (1982) Brain neurotransmitter receptors in Parkinson's disease. In: Marsden CD, Fahn S (eds) Neurology 2: movement disorders. Butterworths, London, pp 59–74

Rinne UK, Koskinen V, Laaksonen H, Löneberg P, Sonninen V (1978) GABA receptor binding in the parkinsonian brain. Life Sci 22:2225–2228

Rinne UK, Koskinen V, Lönnberg P (1980) Neurotransmitter receptors in the parkinsonian brain. In: Rinne UK, Klingler M, Stamm G (eds) Parkinson's disease. Elsevier, Amsterdam, pp 93–107

Rinne UK, Lönnberg P, Koskinen V (1981) Dopamine receptors in the parkinsonian brain. J Neural Transm 51:97–106

Sadoul JL, Checler F, Kitabgi P, Rostene W, Javoy-Agid F, Vincent JP (1984) Loss of high affinity neurotensin receptors in substantia nigra from parkinsonian subjects. Biochem Biophys Res Commun 125:395–404

Shibuya M (1979) Dopamine-sensitive adenylate cylase activity in the striatum in Parkinson's disease. J Neural Transm 44:287–295

Studler JM, Javoy-Agid F, Cesselin F, Legrand JC, Agid Y (1982) CCK-8-Immuroreactivity distribution in human brain: selective decrease in the substantia nigra from parkinsonian patients. Brain Res 243:176–179

Taquet H, Javoy-Agid F, Cesselin F, Hamon M, Legrand JC, Agid Y (1982) Microtopagraphy of methionine-enkephalin, dopamine and noradrenaline in the ventral messencephalon of human control and parkinsonian brains. Brain Res 235:303–314

Taquet H, Javoy-Agid F, Hanon M, Legrand JC, Agid Y, Cesselin F (1983) Parkinson's disease affects differently Met- and Leu-enkephaline in the human brain. Brain Res 280:379–382

Teychenne PF, Bergsrud D, Racy A, Vern B (1981) Low dose bromocriptine therapy in Parkinson's disease. Res Clin Forums 3:37–47

Uhl GR, Whitehouse PJ, Poice DL, Tourtelotte WW, Kuhar MJ (1984) Parkinson's disease: depletion of substantia nigra neurotensin receptors. Brain Res 308:186–190

Correspondence: Dr. N. Ogawa, Department of Neurochemistry, Institute for Neurobiology, Okayama University Medical School, 2-5-1 Shikatacho, Okayama 700, Japan.

Second messengers in Parkinson's disease

N. Nishino[2], N. Kitamura[2], T. Hashimoto[1,2], O. Komure[3], and C. Tanaka[1]

Departments of [1] Pharmacology, and [2] Psychiatry and Neurology, Kobe University School of Medicine, Kobe, and [3] Department of Neurology, Utano National Hospital, Kyoto, Japan

Summary

The second messenger systems coupled to phosphoinositide turnover, diacylglycerol (DG)-protein kinase C (PKC) and inositol 1,4,5-trisphosphate (IP_3)-Ca^{2+} systems, were studied in the striatum of postmortem brains from 20 patients with Parkinson's disease (PD) (Yahr III, IV and V), using ^3H-4β phorbol 12,13-dibutyrate (^3H-PDBu) and ^3H-IP_3 as the respective ligands. In the PD striatum, the specific bindings for ^3H-PDBu and ^3H-IP_3 were not altered when all the patients without dementia, Yahr (III + IV) patients or Yahr V patients, without dementia were compared with control subjects. However, they were significantly decreased in the Yahr V patients with dementia (PDD) by 38% and 78%, respectively. These results suggest that the second messenger sites coupled to the phosphoinositide system are not concentrated in the nigrostriatal dopaminergic neurons and that striatal pathophysiology of PDD is different from that of PD without dementia.

Introduction

No water-soluble signaling molecules including neurotransmitters, hormones and growth factors permeate the plasma membrane. Instead, they bind to specific receptor proteins on the surface of the target cells, which, in turn, transduce the extracellular signals into one or more intracellular signals that alter the function of the target cells. Stimulation of membrane receptors by several neurotransmitters modulates the activities of adenylate cyclase (AC) and phospholipase C (PLC) to yield cAMP, and diacylglycerol (DG) and inositol 1,4,5-trisphosphate (IP_3), respectively. Changes in the amount of these second messengers then modulate the activity of various protein kinases such as cAMP-dependent protein kinase (PKA) and Ca^{2+}/phospholipid-dependent protein kinase (PKC), which, in turn, phosphorylate a series of substrate proteins

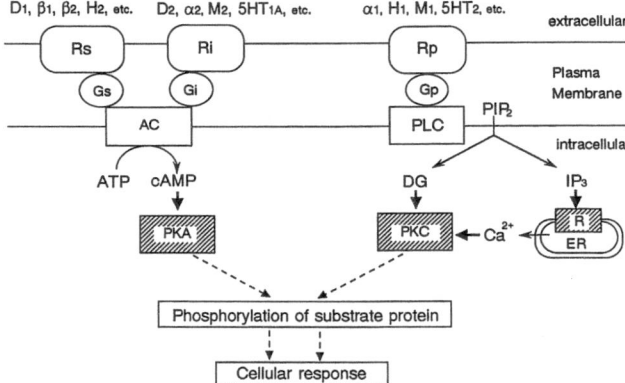

Fig. 1. Transmembrane signalling system. The figure illustrates a sheme for two major signal transduction systems: cAMP system and phosphoinositides system. In the plasma membrane, there are three distinct classes of functional proteins coupled to the signalling systems: receptor proteins, guanine uncleotide-binding proteins (G-proteins) functioning as a transducer, and effector proteins such as adenylate cyclase (AC) and phospholipase C (PLC). For example, stimulation of receptors (R_s) such as dopamine D_1, β-adrenergic and histamine H_2 receptors enables a stimulatory G-protein (G_s), a heterotrimer consisting of α_s, β and γ subunits, to dissociate from the receptor and stimulate an enzyme adenylate cyclase (AC). The activation of AC enhances synthesis of cAMP from ATP. cAMP activates a cAMP-dependent protein kinase (PKA), which then phosphorylates a series of substrate proteins that regulate diverse cellular response. Most transmitters act through both the cAMP and phosphoinositides systems. The discrimination comes in the receptor subtypes involved. For instance, norepinephrine stimulates adenylate cyclase through β-adrenoceptors, but inhibits the enzyme through α_2-adrenoceptor, and stimulates the phosphoinositide cycle through α_1-adrenoceptors. G_i inhibitory G-protein, G_p G-protein coupled to phosphoinositides system, PLC phospholipase C, PIP_2 phosphatidylinositol 4,5-bisphosphate, DG diacylglycerol, IP_3 inositol 1,4,5-trisphosphate, ER endoplasmic reticulum

that regulate subsequent neural activities (Fig. 1). PKC consists of a single polypeptide chain, which possesses regulatory and catalytic domains in its molecule. The regulatory domain of PKC can be labeled by ^{3}H-4β phorbol 12,13-dibutyrate (^{3}H-PDBu) (Nishino et al., 1989), one of the potent tumor-promoting phorbol esters which mimic the action of DG. IP_3 binds to unique receptors on the endoplasmic reticulum and mobilizes Ca^{2+} from the storage sites. IP_3 receptors can be labeled by ^{3}H-IP_3 (Kitamura et al., 1989). The neurochemical pathology of Parkinson's disease (PD) has been clarified by measurements of biogenic amines and their metabolites, and of neurotransmitter receptor binding activities (Birkmayer and Riederer, 1985) (neurotransmitter receptor alterations in PD brains are summarized in Table 1). Investigation of these second messenger systems is necessary to understand how neurotransmitter receptor alterations result in abnormalities in neural functions in neuropsychiatric disorders. We measured the amount of PKC and IP_3 receptor in the striatum of postmortem PD brains, using ^{3}H-PDBu and ^{3}H-IP_3 as respective ligands.

Results and discussion

We have previously reported decreases in the bindings of ^{3}H-PDBu and ^{3}H-IP$_3$ in the caudate nucleus, putamen and globus pallidus but not in the frontal cortex of PD patients of Yahr V, and suggested that derangements of striato-nigral and -pallidal neurons such as GABA-, substance P-, and enkephalin-containing neurons are related to the decreased bindings (Nishino et al., 1989; Kitamura et al., 1989). In this study, we included PD patients who had been diagnosed antemortem as Yahr III and IV as well as Yahr V and those cases that had been complicated with dementia, and reexamined these binding activities in a bigger sample size.

^{3}H-PDBu and ^{3}H-IP$_3$ binding activities were unaltered in the striatum of all patients with PD (Yahr III + IV + V). The results appear to support the view that the striatal efferent neurons are intact in PD. However, when the patients were subdivided based on Yahr's criteria of clinical stage (Hoehn and Yahr, 1967), and on the presence and absence of concomitant dementia, some non-negligible discrepancies among the subgroups became evident. Although not statistically significant, there is a tendency for the binding activities to increase in the striatum of Yahr (III + IV) patients by 25% and 37%, respectively. The specific bindings of ^{3}H-PDBu and ^{3}H-IP$_3$ were significantly decreased in the striatum of PDD by 38% and 78%, respectively, but remained unchanged in the striatum of PD patients without dementia (Table 2).

These observations suggest that pathophysiological heterogeneity exists in PD with respect to phosphoinositide systems in that the severest loss of these second messenger sites seems to occur in the striatum of PDD, which may be attributable to the possible loss of striatal neurons and/or striatal afferent neurons. In fact, morphometric analysis of the brain slices from PDD patients has revealed a 38% loss of tissue in the putamen and globus pallidus combined (De La Monte et al., 1989), which may be comparable to the decrease of ^{3}H-PDBu binding in the PDD striatum in this study. Also, a severer neuronal loss has been shown in the locus coeruleus as well as in the nucleus basalis of Meynert of PDD patients than in that of PD patients without dementia (Gaspar and Gray, 1984). The unaltered bindings of ^{3}H-PDBu and ^{3}H-IP$_3$ in the striatum of PD further suggest that PKC and IP$_3$ receptors abound in the striatal efferent neurons such as GABA-, substance P- and enkephaline-containing neurons in the striatum rather than in the nigro-striatal dopaminergic neural pathway, as demonstrated by lesion studies of the rat brain (Worley et al., 1986).

Recent cDNA cloning has revealed that there are at least 7 subspecies (a, βI, βII, γ, ε, δ, ζ) of PKC (Nishizuka, 1988). Immunocytochemical and lesion studies showed a heterogenous localization of the PKC subspecies in the striatum and substantia nigra, suggesting the involvement of the a-PKC in the function of the nigro-striatal dopaminergic neurons and striatal intrinsic cholinergic neurons, the βI-PKC in the function of the striatal intrinsic GABAergic neurons, and the βII- and γ-PKCs in the function of the striatonigral neurons (Yoshihara et al., 1991). Although ^{3}H-PDBu can label at least a, βI, βII, and

Table 1. Neurotransmitter receptor alterations in Parkinson's disease

Receptors (subclass)	Radioligands	Alterations	Reporters (first author)
Dopamine			
(D_1)	^{3}H-flupenthixol	SN, PUT: L-DOPA($-$) $\downarrow$ L-DOPA($+$) $\uparrow$	Rinne (1985)
	^{3}H-SCH-23390	PUT $\rightarrow$	Pimoule (1985)
		SNC $\downarrow$; SNR, PUT $\rightarrow$	Cash (1987)
(D_2)	^{3}H-spiperone	PUT: L-dopa($-$) $\uparrow$ L-dopa($+$) $\rightarrow$ $\downarrow$	Rinne (1981)
		PUT $\uparrow$; ACC, CN, SN $\rightarrow$	Bokobza (1984)
		PUT $\uparrow$; ACC $\rightarrow$	Javoy-Agid (1984)
		CN, PUT $\rightarrow$	Guttman (1986)
		PUT $\rightarrow$	Cash (1987)
		PUT $\downarrow$	Reisine (1977)
		PUT $\downarrow$	Winkler (1980)
	^{3}H-haloperidol	PUT $\uparrow$	Lee (1978)
Acetylcholine			
($m_1 + m_2$)	^{3}H-QNB	PUT $\uparrow$; CN, PAL $\rightarrow$	Reisine (1977)
		CN, FC $\uparrow$	Ruberg (1982)
		CN, PUT, PAL, CN, FC $\rightarrow$	Nishino (1988)
(m_1)	^{3}H-pirenzepine	PUT $\rightarrow$	Nishino (1988)
(nicotinic)	^{3}H-acetylcholine	FC $\downarrow$	Whitehouse (1988)
GABA-BDZ			
(GABA)	^{3}H-GABA	SN $\downarrow$; CN, PUT $\rightarrow$	Lloyd (1977) Reisine (1977) Rinne (1978)
(GABA$_A$)	^{3}H-muscimol	CN, PUT, PAL, SN, FC $\downarrow$	Nishino (1988)
(BDZ)	^{3}H-flunitrazepam	CN $\uparrow$; PUT, FC $\rightarrow$	Maloteaux (1988)
(BDZ type I)	^{3}H-flunitrazepam	SN $\downarrow$	Uhl (1986)
(BDZ type II)		SN $\rightarrow$	Uhl (1986)
Glutamate			
(NMDA)	^{3}H-TCP	HIP, FC $\rightarrow$	Shirakawa (1989)
Glycine			
	^{3}H-strychinine	SN $\downarrow$	Lloyd (1983)
Norepinephrine			
(a_1)	^{3}H-prazocine	FC $\uparrow$	Cash (1984)
(a_2)	^{3}H-clonidine	FC $\downarrow$	Cash (1984)
	^{3}H-rauwolscine	LC $\rightarrow$	Cash (1987)
(β_1)	^{3}H-DHA	FC $\uparrow$	Cash (1984)
(β_2)	^{3}H-DHA	FC $\rightarrow$	Cash (1984)
Serotonin			
(5-HT$_1$)	^{3}H-serotonin	PUT $\downarrow$	Reisine (1977)
		FC $\downarrow$	Kienzl (1981)
		FC, CN, PUT $\rightarrow$	Rinne (1980)
(5-HT$_{1A}$)	^{3}H-80H-DPAT	FC $\rightarrow$	Hashimoto (1991)
(5-HT$_2$)	^{3}H-spiperone	SN $\rightarrow$	Uhl (1986)

Table 1. Continued

Receptors (subclass)	Radioligands	Alterations	Reporters (first author)
Histamine (H₁)	³H-mepyramine	FC →	Nakai (1988)
Somatostatin	¹²⁵I-LTT-SS-28	SN →	Uhl (1986)
Neurotensin	³H-neurotensin	SN ↓	Uhl (1986)
Opioid	³H-etorphine	CN,PUT,PAL,SNC,AMG,FC →	Delay-Goyet (1987)
(μ)	³H-naloxone	CN ↓	Reisine (1979)
	³H-dihydromorphine	SN ↓	Uhl (1986)
	³H-DAGO	CN,PUT,PAL,SNC,AMG,FC →	Delay-Goyet (1987)
(κ)	³H-EKCZ	SN ↓	Uhl (1986)
(δ)	³H-DTLET	CN,PUT,PAL,AMG,FC →	Delay-Goyet (1987)
	³H-leu-enkephalin	PUT,HIP,LIM ↑	Rinne (1983)
	³H-met-enkephalin	CN,PUT,ACC,HIP,FC ↑	Rinne (1983)
Dopamine uptake site	³H-cocaine	PUT ↓	Shoemarker (1985)
	³H-GBR-12935	CN ↓	Janowsky (1987)
	³H-GBR-12935	PUT ↓	Shirakawa (1988)
5-HT uptake site	³H-imipramine	PUT,PFC ↓	Cash (1985)
		PUT ↓	Raisman (1986)
	³H-paroxetine	PUT ↓	Raisman (1986)
Norepinephrine uptake site	³H-DMI	LC →	Cash (1987)
Glutamate uptake site	³H-D-aspartate	CX →	Cross (1987)
Ca²⁺ channel	³H-nitrendipine	CN,PUT,SN ↓	Nishino (1986)
	³H-(+)-PN200-110	CN + PUT →	Watson (1988)

→ unaltered; ↑ increased; ↓ decreased

SNC(R) substantia nigra pars compacta (pars reticulata), *CN* caudate nucleus, *PUT* putamen, *PAL* globus pallidus, *FC* frontal cortex, *PFC* prefrontal cortex, *LIM* limbic area, *HIP* hippocampus, *AMG* amygdala, *ACC* nucleus accumbens, *GABA* γ-aminobutyric acid, *BDZ* benzodiazepine, *5-HT* 5-hydroxytryptamine, *ACE* angiotensin converting enzyme, *QNB* quinuclidinyl benzilate, *DHA* dihydroalprenolol, *TCP* 1-1-(2-thienyl)cyclohexylpiperidine, *LTT-SS-28* Leu-8-dTrp-Tyr-SS-28, *DAG0* Tyr-D-Ala-Gly-(Me)Phe-Gly-ol, *EKCZ* ethylketocyclazocine, *DTLET* Tyr-D-Thr-Gly-Phe-Leu-Thr, *DMI* desmethylimipramine

Table 2. *Binding activities of ^3H-PDBu and ^3H-IP$_3$ in the striatum of patients with Parkinson's disease.* Brains were obtained at autopsy at the Department of Anatomy, Kobe University School of Medicine (Kobe) and the Department of Neurology, Utano National Hospital (Kyoto) from 20 patients with idiopathic PD and 24 control subjects with no known neuropsychiatric disorders. There was no statistically significant difference between these two groups with respect to age or interval from death to autopsy. Based on the criteria of Hoehn and Yahr (4), three patients had been diagnosed antemortem as Yahr III, two patients as Yahr IV, and fifteen patients as Yahr V. Seven of fifteen Yahr V patients had been complicated with dementia. Radioreceptor assay was performed by rapid filtration technique as described previously (Nishino et al., 1989; Kitamura et al., 1989). For protein kinase C monitoring, the membrane was incubated with ^3H-PDBu (4.1 nM) in 50 mM Tris-HCl buffer (pH 7.4 at 25°C) containing 4 mg/ml of bovine serum albumin for 30 min at 30°C. For IP$_3$ receptor-binding, the membrane was incubated with ^3H-IP$_3$ (7.1 nM) in 50 mM Tris-HCl buffer (pH 8.3 at 25°C) containing 1 mM EDTA and 1 mg/ml of bovine serum albumin for 10 min at 4°C. The non-specific bindings were determined in the presence of 1 μM PDBu and 1 μM IP$_3$, respectively

| | Specific binding (fmol/mg protein, mean ± SEM) | |
	^3H-PDBu	^3H-IP$_3$
Control	2893 ± 203	120 ± 12
	(n = 24)	(n = 24)
PD	2729 ± 222	114 ± 17
	(n = 20)	(n = 16)
Yahr (III + IV)	3604 ± 335	164 ± 14
	(n = 5)	(n = 5)
Yahr V	2999 ± 218	146 ± 17
without	(n = 8)	(n = 6)
Dementia		
Yahr V with	1797 ± 282*§	27 ± 11**§
Dementia	(n = 7)	(n = 5)

Significant difference from means of controls with: * P < 0.05, ** P < 0.01
Significant difference from means of Yahr V without dementia with: § P < 0.005

γ-PKCs, the decreased ^3H-PDBu bindings in the striatum of PDD may reflect the loss of βII-PKC immunoreactive neurons, which are probably GABAergic, since most of the βII- and γ-PKC immunoreactive neurons were medium-sized neurons projecting to the substantia nigra, and over 90% of GABAergic neurons in the striatum contained βII-PKC (Yoshihara et al., 1991). It remains to be studied whether or not the phosphorylation of a certain substrate protein for PKC is altered in a particular brain area of PD. The recent increased knowledge of the intracellular signal transduction system suggests this area of PD research to be an increasingly attractive possibility of new therapeutic approach, and makes it an important direction of future studies.

Acknowledgments

We thank Prof. T. Yamadori, Department of Anatomy, Kobe University School of Medicine, and Dr. S. A. Noguchi-Kuno, Department of Neurology, Utano National Hospital, for provinding us with valuable autopsied brain specimens. This work was supported by a Grant-in-Aid for Special Project Research of Selected Intractable Neurological Disorders from the Ministry of Education, Science and Culture, Japan.

References

Birkmayer W, Riederer P (1985) Die Parkinson-Krankheit. Biochemie, Klinik, Therapie, 2. Aufl. Springer, Wien New York

De La Monte SM, Wells SE, Hedley-Whyte ET, Growdon JH (1989) Neuropathological distinction between Parkinson's dementia and Parkinson's plus Alzheimer's disease. Ann Neurol 26:309–320

Gaspar P, Gray F (1984) Dementia in idiopathic Parkinson's disease: a neuropathological study of 32 cases. Acta Neuropathol (Berl) 64:43–52

Hoehn MM, Yahr MD (1967) Parkinsonism: onset, progression and mortality. Neurology 17:427–442

Kitamura N, Hashimoto T, Nishino N, Tanaka C (1989) Inositol 1,4,5-trisphosphate binding sites in the brain: regional distribution, characterization, and alterations in brain of patients with Parkinson's disease. J Mol Neurosci 1:181–187

Nishino N, Kitamura N, Nakai T, Hashimoto T, Tanaka C (1989) Phorbol ester binding sites in human brain: characterization, regional distribution, age-correlation, and alterations in Parkinson's disease. J Mol Neurosci 1:19–26

Nishizuka Y (1988) The molecular heterogeneity of protein kinase C and its implications for cellular regulation. Nature 334:661–665

Worley PF, Baraban JM, Snyder SH (1986) Heterogenous localization of protein kinase C in rat brain: autoradiographic analysis of phorbol ester receptor binding. J Neurosci 6:199–207

Yoshihara C, Saito N, Taniyama K, Tanaka C (1991) Differential localization of four subspecies of protein kinase C in the rat striatum and substantia nigra. J Neurosci (in press)

Correspondence: Prof. C. Tanaka, Kobe University School of Medicine, 7-5-1 Kusunoki-cho, Chuo-ku, Kobe 650, Japan.

Mitochondrial energy crisis as a mechanism for nigral cell death

Y. Mizuno

Department of Neurology, Juntendo University School of Medicine, Tokyo, Japan

Summary

Studies on mitochondrial respiratory enzymes in Parkinson's disease are reviewed. The presence of abnormalities in Complex I of the mitochondrial respiratory chain was described in Parkinson's disease by several groups of investigators utilizing different methods. These abnormalities do not seem to be a primary defect of Parkinson's disease. Instead, they seem to be a consequence of long-standing noxious stimuli to substantia nigra. Dopaminergic neurons in substantia nigra appear to have a unique property with reduced tolerance to various noxious stimuli. Further studies on mitochondria seem to be important for the elucidation of Parkinson's disease.

Introduction

Etiology of Parkinson's disease is still unknown; however, the discovery of the nigral neurotoxin, 1-methyl-4-phenyl-1,2,3,6-tetrahydropyridine (MPTP), has made a great contribution to the research on Parkinson's disease, including its etiology. Several groups of investigators found mitochondrial respiratory failure in this model (Nicklas et al., 1985; Heikkila et al., 1985; Ramsay et al., 1986; Mizuno et al., 1987a). Stimulated by these studies, mitochondrial functions have also been studied in Parkinson's disease, and some interesting results have been reported (Schapira et al., 1989, 1990; Bindoff et al., 1989; Mizuno et al., 1989, 1990; Parker et al., 1990). Deleterious effects of mitochondrial abnormalities on the cellular integrity have been well demonstrated in mitochondrial encephalomyopathies. In Parkinson's disease, studies on mitochondria began only a few years ago. In this article, I will briefly review the fundamental functions of mitochondria, and present the recent data on mitochondria of Parkinson's disease.

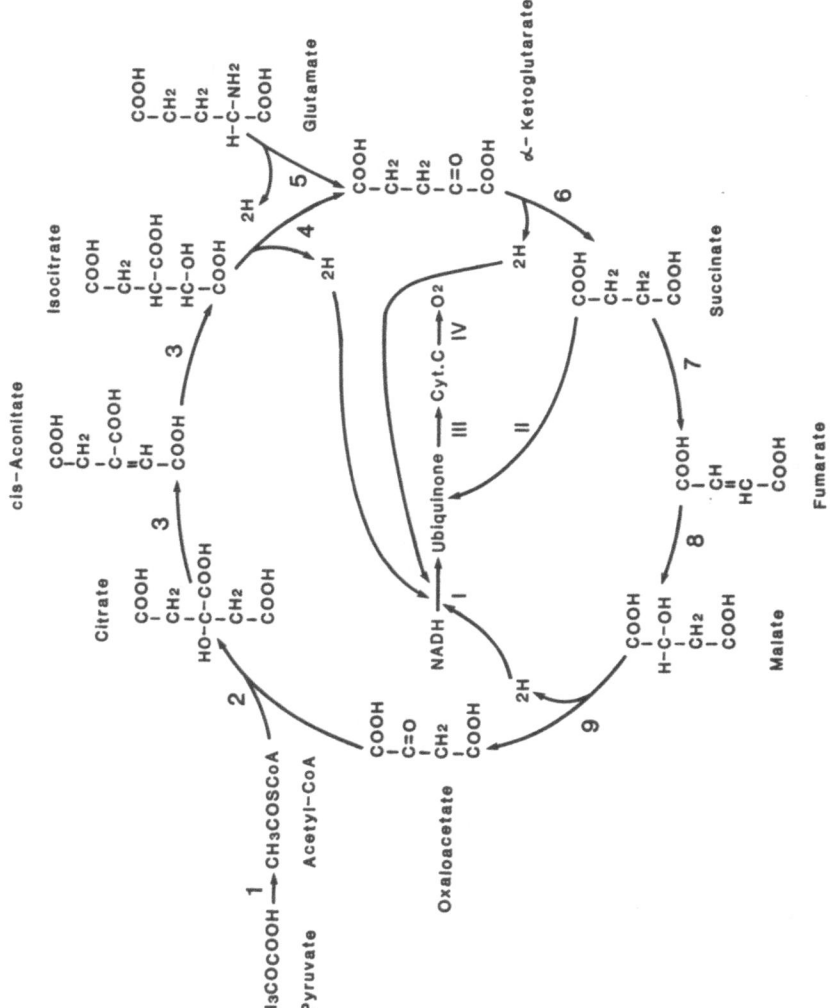

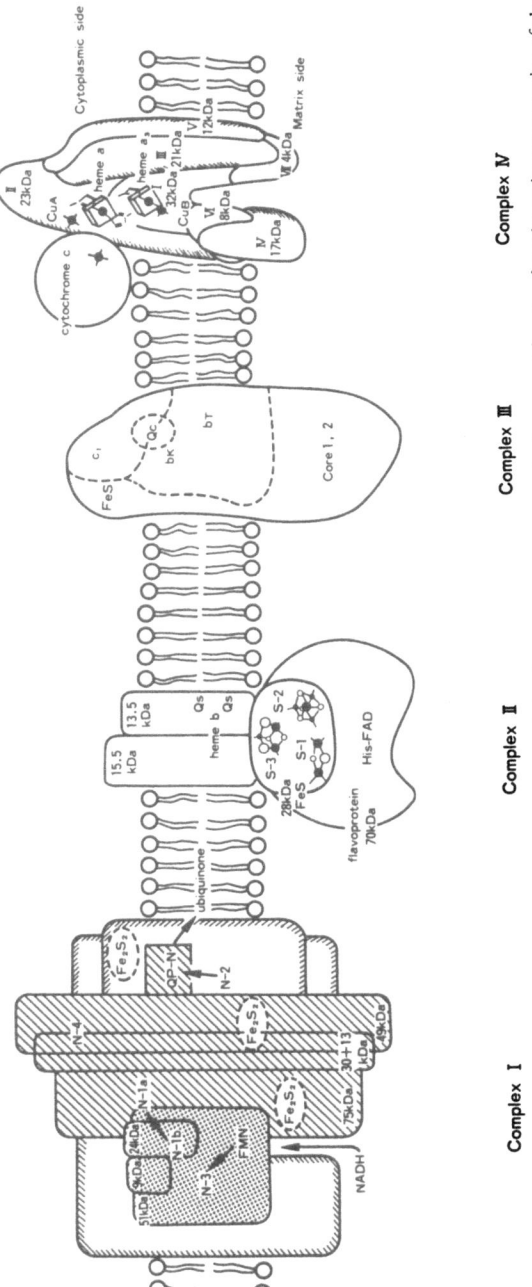

Complex I   Complex II   Complex III   Complex IV

Fig. 1. A schematic representation of the tricarboxylic acid cycle and the electron transport system in the center (above) and a schematic representation of the enzyme-protein complexes of the electron transport system (below). The latter was adopted from the schema by Dr. T. Ozawa of Department of Biomedical Chemistry, University of Nagoya

Mitochondria

Mitochondria are intracellular organella of about 0.4 to 4 micrometers in length. They are extremely important for cellular respiration and the production of high-energy phosphate bonds of adenosine triphosphate (ATP). Oxygen utilization and ATP synthesis are achieved by the electron transport system located in the inner membrane of mitochondria, as schematically shown in Fig. 1. The tricarboxylic acid (TCA)-cycle (Fig. 1) supplies NADH and succinate as substrates for the electron transport system from pyruvate, the final product of the glycolytic pathway. The electron transport system consists of 4 protein-enzyme complexes, i.e., Complexes I, II, III and IV as schematically shown in Fig. 1. Electrons are transported from NADH to ubiquinone by Complex I, from uniquinol to cytochrome c by Complex III, and finally from reduced cytochrome c to oxygen by Complex IV. Electrons are also transported from succinate to ubiquinone by Complex II. At each step of electron transfer except for Complex II, ATP is formed from ADP by the action of Complex V.

When the efficiency of the electron transport system decreases, not only the amount of ATP formed is reduced, but also the complete reduction of oxygen molecules becomes impaired giving rise to oxygen free radicals (Linnane et al., 1989), which are very toxic to living cells through by various mechanisms (Halliwell, 1989). ATP is essential for the maintenance of cellular integrity and the specialized functions of neurons. For instance, ATP is necessary for Na^+-K^+ ATPase, biosynthesis of neurotransmitters, axonal transport, neural transmissions, cyclic nucleotide synthesis, protein phosphorylation by protein kinases, biosynthesis of nucleic acids and glycolysis, among others. Therefore, inhibition of mitochondrial respiration and the resultant decrease in ATP formation will cause tremendous deleterious effects on neurons.

Some of the subunits of the electron transport complexes are coded by mitochondrial DNA. Mitochondrial DNA also codes two ribosomal RNAs and transfer RNAs for 22 amino acids necessary for protein synthesis within mitochondria. Mitochondrial DNA is more vulnerable to noxious stimuli compared to nuclear DNA because of a higher frequency of mutations and the lack of repair mechanisms (Richter et al., 1988; Linnane et al., 1989). Therefore, long-standing noxious stimuli may produce mitochondrial DNA damage with resultant defects in the electron transport system. Mitochondrial DNA deletion or mutation is the cause of cellular damage in mitochondrial encephalo-myopathies (Holt et al., 1988; Ozawa et al., 1988; Moraes et al., 1989; Tanaka et al., 1989; Schofner et al., 1990). Most of them are sporadic cases, and it is interesting to note that such sporadic disorders have defects in DNA. A similar mechanism may be operating locally in Parkinson's disease. A hint for the mitochondrial hypothesis resulted from the study on MPTP.

MPTP-induced parkinsonism

MPTP was found as a by-product of a homemade narcotic (Davis et al., 1979). An American youngster who injected MPTP suffered from severe parkinsonism. He eventually died of drug overdose. His substantia nigra showed a marked degeneration. In 1983, Langston et al. (1983) reported similar patients in California, which triggered intense interest in MPTP. Soon, MPTP was found to be oxidized by monoamine oxidase B to MPP^+ (Chiba et al., 1984a), and MPP^+ turned out to be the toxic species against substantia nigra (Heikkila et al., 1984; Irwin and Langston, 1985). Concerning the mechanism of selective destruction of substantia nigra, it was found that MPP^+ was actively taken up into nigro-striatal dopaminergic neurons (Chiba et al., 1984b; Javitch and Snyder, 1984), and concentrated within them. Neuromelanin also appears to play a role in accumulating MPP^+ within nigra (Lyden et al., 1983).

With regard to the mechanism of neuronal degeneration in this model, Ramsay and Singer (1986) found an energy-dependent active transport system for MPP^+ in mitochondria. As mitochondria begin to respire, the interior of the mitochondria becomes negatively charged as positively charged protons (H^+) are expelled out of the mitochondria. Positively charged MPP^+ are concentrated within mitochondria according to the charge gradient.

Inhibition of mitochondrial respiration by MPP^+ was reported independently by 3 groups (Nicklas et al., 1985; Ramsay et al., 1986; Mizuno et al., 1987a). Table 1 summarizes our data on the effects of MPP^+ on mitochondrial respiration. We observed inhibition of mitochondrial Complex I by MPTP and MPP^+ to be the origin of mitochondrial respiratory dysfunction (Mizuno et al., 1986, 1987b). Alpha-ketoglutarate dehydrogenase complex in the TCA cycle is also inhibited by MPP^+ (Mizuno et al., 1987c). As the latter enzyme is the rate-limiting enzyme in the TCA cycle, its inhibition would cause an additional

Table 1. Effect of MPP^+ on mitochondrial respiration and respiratory enzymes

	No	Control	MPP^+ (Concentration)	P
State 3 respiration	9	180 ± 25	21 ± 4 (0.05)	***
State 4 respiration	9	38 ± 5	11 ± 3 (0.05)	***
Respiratory quotient	9	4.7 ± 0.8	2.0 ± 0.3 (0.05)	***
ADP/O	9	2.6 ± 0.2	not calculated	
ATP formed#	4	637 ± 23	210 ± 31 (0.05)	***
Complex I (fresh)	9	57 ± 13	27 ± 6 (0.05)	***
Complex I (freeze-thawed)	9	247 ± 52	151 ± 38 (0.05)	***
Glutamate dehydrogenase	8	556 ± 50	545 ± 52 (2.0)	
Isocitrate dehydrogenase	8	285 ± 37	266 ± 35 (2.0)	
Malate dehydrogenase	8	4760 ± 530	4870 ± 680 (2.0)	
α-ketoglutarate dehydrogenase	8	34 ± 5	19 ± 3 (2.0)	***

Mean $\pm$ SE, Enzyme source = mouse brain mitochondria, units: State 3 and state 4 = nanoatom 0 utilized/min/mg protein, Enzymes = nanomole substrate processed/mg protein, # = ATP formed from 750 nanomoles of ADP, fresh = activity assayed using intact mitochondria, freeze-thawed = activity assayed after freeze-thawing of mitochondria, Dehydrogenases are NAD^+-linked enzymes. Numbers in parentheses indicate concentration of MPP^+ (mM). P: P values, non-paired t-Test, *** $P < 0.001$, from reference Mizuno et al. (1988)

Table 2. Mitochondrial enzymes in Parkinson's disease

	Tissue	Control	Parkinson	P	Reference
Complex I	SN	3.36 ± 0.44 (5)	2.34 ± 0.76 (5)	**	S
	Str	43 ± 34 (5)	33 ± 22 (5)		M
	Plt	19.1 ± 5.6 (8)	8.7 ± 2.9 (8)	***	P
	Mus	222 ± 30 (4)	132 ± 40 (5)	***	B
Complex I + III	SN	4.36 ± 1.41 (9)	2.68 ± 1.01 (9)	***	S
Complex II	Str	83 ± 28 (5)	59 ± 34 (5)		M
	Mus	353 ± 70 (4)	181 ± 33 (5)	***	B
Complex II + III	SN	9.46 ± 3.01 (9)	9.59 ± 3.15 (9)		S
	Plt	225 ± 98 (8)	155 ± 70 (10)		P
Complex III	Str	206 ± 45 (5)	133 ± 50 (5)	*	M
Complex IV	Str	70 ± 20 (5)	50 ± 29 (5)		M
	Plt	167 ± 36 (4)	155 ± 73 (5)		P
	Mus#	2.05 ± 0.23 (4)	1.23 ± 0.31 (5)	***	B

Tissue: *SN* substantia nigra, *Str* striatum, *Plt* platelet, *Mus* skeletal muscle, P: P values, *** $P < 0.01$, ** $P < 0.02$, * $P < 0.05$, Mean ± SD, Numbers in parentheses indicate the number of patients. Unit = nanomole substrate processed/minute/mg protein, # = expressed as apparent first-order rate constant, k/s per mg protein. References: S = Schapira et al. (1990); M = Mizuno et al. (1990); P = Parker et al. (1990); B = Bindoff et al. (1989)

damage to mitochondrial respiration. Without operation of the TCA cycle, NADH cannot be supplied to the electron transport system.

Thus, it seems to be well accepted that the mechanism of neuronal degeneration in MPTP-induced parkinsonism is the energy crisis due to the inhibition of mitochondrial respiration. It is amazing that an exogenous substance such as MPP^+ is selectively and highly concentrated within nigral mitochondria. Suppose a similar toxic molecule is present in the environment or is formed within the human body; it may eventually damage nigral neurons of susceptible persons; leading to Parkinson's disease.

Mitochondria in Parkinson's disease

Stimulated by the research on MPTP-induced parkinsonism, several groups of investigators studied mitochondria in Parkinson's disease. Most of them reported a decrease in the activity of Complex I (Table 2). We found a decrease in Complex III activity in the striata of patients with Parkinson's disease. The assay of activities of the mitochondrial electron transport system in human autopsy specimens has several problems. The specific activities assayed in autopsy materials are very low compared with those assayed using fresh animal brains. The activities of most of the mitochondrial respiratory enzymes are reasonably stable between the 4-to-12-hour postmortem period, except for Complex I (Mizuno et al., 1990). The very low specific activity appears to be

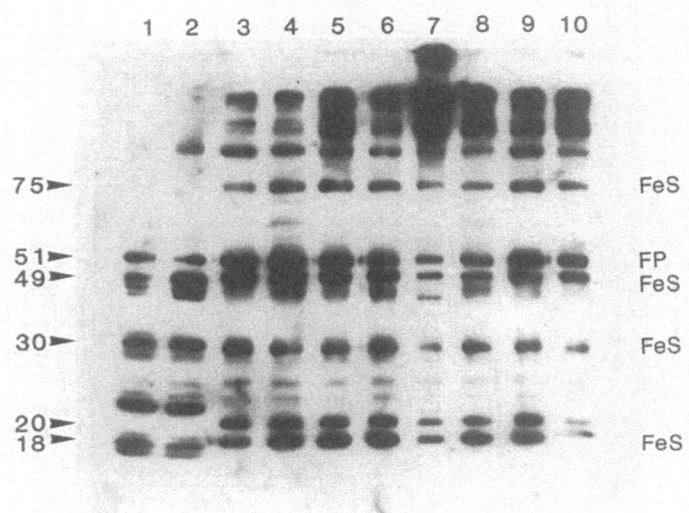

Fig. 2. Immunoblot analysis of the subunits of Complex I. *Lane 1* purified Complex I; *Lane 2* bovine heart mitochondria; *Lane 3–5* control; *Lane 6–10* patients with Parkinson's disease. In lane 7 to 10, stainings of the four bands between 24 and 33kDa subunits are moderately to markedly diminished. From reference Mizuno et al. (1989)

due to either the loss of activity during the very early postmortem period or during the premortem agonal stage. Therefore, due care is necessary to form a conclusion from the studies on human brains.

To complement this drawback, we also studied the subunits of the electron transport system by Western blot analysis and mitochondrial DNA by the polymerase chain reaction (PCR) method. The details of the methods are described elsewhere (Ikebe et al., 1990).

Figure 2 represents the Western blot analysis of Complex I using striatal mitochondria of patients with Parkinson's disease. In 4 out of the 5 patients with Parkinson's disease, the stainings of subunits corresponding to 24 to 33-kDa subunits of Complex I were moderately reduced. Also, we studied substantia nigra by immunohistochemical methods for Complexes I, II, III and IV. We found nigral melanin containing neurons with reduced staining against Complex I antibody in all 8 cases of Parkinson's disease studied (manuscript in preparation). When the midbrain sections were stained for Complexes III and IV, practically all of the neurons were stained well. When they were stained with antiserum against Complex II, 3 patients with Parkinson's disease showed reduced staining. In control patients, stainings against antisera Complexes II, III and IV were uniformly good. When the midbrain sections of control patients were stained for Complex I, some nigral neurons showed reduced staining; however, the proportion of neurons with reduced staining was significantly

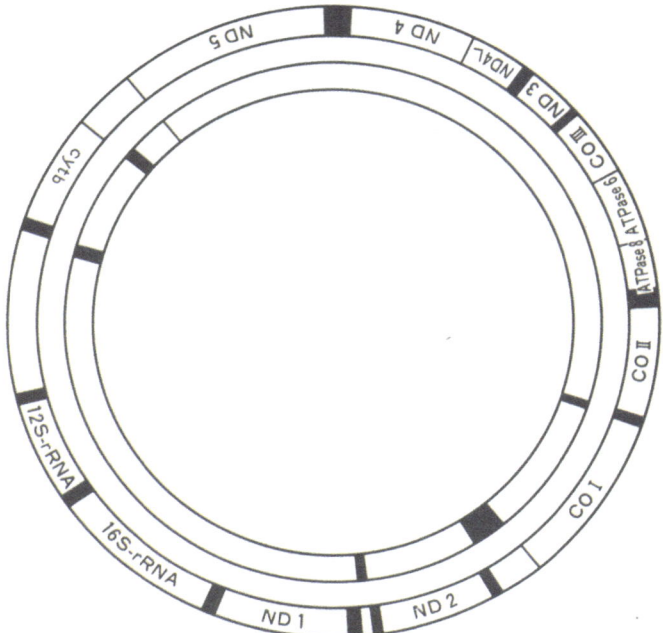

Fig. 3. A schematic representation of mitochondrial DNA and its coding regions. The black areas represent the coding regions for ribosomal RNAs

smaller than that of Parkinson's disease. These findings prompted us to investigate mitochondrial DNA in Parkinson's disease.

Mitochondria have a circular DNA consisting of 16,569 base pairs (Fig. 3) (Anderson et al., 1981). In mitochondrial myopathies, various kinds of partial deletions have been reported (Holt et al., 1988; Ozawa et al., 1988; Moraes et al., 1989; Tanaka et al., 1989). Among them, the deletion of about 5000 base pairs between the ATPase 8 gene and the ND 5 (NADH dehydrogenase subunit 5) gene is very frequent (Tanaka et al., 1989). Therefore, we studied mitochondrial DNA from striata of patients with Parkinson's disease by the PCR method using synthetic oligoprimers which encompass the ATPase 8 gene and the ND 5 gene (Ikebe et al., 1990). We found a 4977 base pair deletion between the 13 base pair direct repeat sequences in those genes in all patients with Parkinson's disease studied. Two elderly control patients also showed the same deletion, but the 5 other younger patients did not show the deletion. In the frontal cortex, the same deletion, but of a much smaller amount, was also found in all parkinsonian patients and the two controls mentioned above. By a semiquantitative analysis, the amount of the deleted mitochondria was estimated to be approximately 5% of the total mitochondrial DNA (Ozawa et al., 1990).

Discussion

Reported mitochondrial abnormalities in Parkinson's disease can be summarized as follows: 1) reduced activity of Complex I in substantia nigra and the skeletal muscle; 2) reduced activity of Complex III in striatum and the skeletal muscle; 3) reduced activity of Complex II in the skeletal muscle; 4) reduction in 4 Complex I subunits in striatum; 5) reduced staining against Complex I antibody in substantia nigra; 6) presence of mitochondria with a mitochondrial DNA deletion in striatum. At present, it is not clear whether these abnormalities represent a primary defect, a secondary effect or a by-chance observation. Nonetheless, these findings suggest the presence of mitochondrial abnormalities in Parkinson's disease. As the magnitude of the loss of Complex I activity is relatively small, it does not seem to be a kind of congenital defect. The magnitude of reduction is much smaller, even when compared with the mildest form of mitochondrial encephalomyopathy with Complex I deficiency (Seyama et al., 1989).

The reduction in the amount of Complex I subunits revealed by the immunoblotting and immunostaining studies suing the antiserum against Complex I is consistent with the reduced activity of Complex I in Parkinson's disease. The mitochondrial DNA deletion that we observed involves ND 4 and ND 5 genes which code respective subunits of Complex I. The molecular weight of subunit 5 of Complex I is approximately 30,000, and our immunoblotting analysis showed reduction in the subunits with the molecular weight between 24 and 33 kDa. Therefore, it seems to be likely that the reduced activity and amount of Complex I are the results of mitochondrial DNA deletion. Reduced activity of Complex I will cause a loss of efficiency in electron transport and ATP production. Incomplete reduction of oxygen molecules may result in the formation of highly toxic oxygen free radicals, such as superoxide anions and hydroxyl radicals, which may also play a role in further damaging mitochondrial membranes, proteins and DNA. Reduced formation of ATP will cause derangements in various important metabolic pathways including the neural transmission.

What, then, might be the cause of mitochondrial deletion? Parkinson's disease is usually sporadic. Even familial Parkinson's disease is usually transmitted as the autosomal dominant or recessive trait. Therefore, a genetic defect in mitochondrial DNA seems unlikely. It may be a cumulative effect of long-standing noxious stimuli, including metabolic changes related to aging, as suggested by Linnane et al. (1989). The fact that aged control patients also showed a similar defect in mitochondrial DNA suggests that observed mitochondrial abnormalities may in part be caused by the aging process. However, aging per se does not produce Parkinson's disease. Some unknown constitutional difference may be present in those who eventually develop Parkinson's disease. The tendency to avoid smoking and the rather modest, reserved and meticulous character buildup of parkinsonian patients may in part be manifestations of a constitutional aberration. A search for genetic bases for such differences seems also to be important.

Acknowledgements

This study was in part supported by the Grand-in-Aid for the Priority Areas from Ministry of Education, Science and Culture, Japan. The antisera against Complexes I and III were generous gifts of Prof. T. Ozawa of the Department of Biomedical Chemistry, University of Nagoya, and the antiserum against Complex II, of Prof. K. Kita of the Department of Parasitology, Juntendo University School of Medicine, and that against Complex IV, of Prof. T. Satoh of the Department of Neurology, Juntendo University School of Medicine. We thank Ms. M. Chinahara for the preparation of the manuscript.

References

Anderson S, Bankier AT, Barrell BG, de Bruijin MHL, Coulson AR, Crouin J, Eperon IC, Nierlich DP, Roe BA, Sanger F, Schreier PH, Smith AJH, Young IG (1981) Sequence and organization of the human mitochondrial genome. Nature 290:457–465

Bindoff LA, Birch-Machin M, Cartlidge NEF, Parker WD Jr, Turnbull DM (1989) Mitochondrial function in Parkinson's disease. Lancet ii:49

Chiba K, Trevor AJ, Castagnoli N Jr (1984a) Metabolism of the neurotoxic tertiary amine, MPTP, by brain monoamine oxidase. Biochem Biophys Res Commun 120:574–578

Chiba K, Trevor AJ, Castagnoli N Jr (1984b) Active uptake of MPP$^+$, a metabolite of MPTP, by brain synaptosomes. Biochem Biophys Res Commun 128:1229–1232

Davis GC, Williams AC, Markey SP, Ebert MH, Caine ED, Reichert CM, Kopin IJ (1979) Chronic Parkinsonism secondary to intravenous injection of meperidine analogues. Psychiatry Res 1:249–254

Halliwell B (1989) Oxidants and the central nervous system: some fundamental questions. Acta Neurol Scand 126:23–33

Heikkila RE, Manzino L, Cabbat FS, Duvoisin RC (1984) Protection against the dopaminergic neurotoxicity of 1-methyl-4-phenyl-1,2,5,6-tetrahydropyridine by monoamine oxidase inhibitors. Nature 311:467–469

Heikkila RE, Nicklas WJ, Vyas I, Duvoisin RC (1985) Dopaminergic toxicity of rotenone and the 1-methyl-4-phenylpyridinium ion after their stereotaxic administration to rats: implication for the mechanism of 1-methyl-4-phenyl-1,2,3,6-tetrahydropyridine toxicity. Neurosci Lett 62:389–394

Holt IJ, Harding AE, Morgan-Hughes JA (1988) Deletions of muscle mitochondrial DNA in patients with mitochondrial myopathies. Nature 331:717–719

Ikebe S, Tanaka M, Ohno K, Sato W, Yamamoto T, Hattori K, Ozawa T, Kondo T, Sato T, Mizuno Y (1990) Increased deleted mitochondrial DNA in the striatum in Parkinson's disease and senescence. Biochem Biophys Res Commun 170:1044–1048

Irwin I, Langston JW (1985) Selective accumulation of MPP$^+$ in the substantia nigra: a key to neurotoxicity? Life Sci 36:207–212

Javitch JA, Snyder SH (1984) Uptake of MPP$^+$ by dopaminergic neurons explains selectivity of parkinsonism inducing neurotoxin MPTP. Eur J Pharmacol 106:455–456

Langston JW, Ballard P, Tetrud JW, Irwin I (1983) Chronic parkinsonism in humans due to a product of meperidine-analog synthesis. Science 219:979–980

Linnane AW, Marzuki S, Ozawa T, Tanaka M (1989) Mitochondrial DNA mutations as an important contributor to ageing and degenerative disease. Lancet i:642–645

Lyden A, Bondesson V, Larsson BS (1983) Melanin affinity of 1-methyl-4-phenyl-1,2,3,6-tetrahydropyridine, an inducer of chronic Parkinsonism in humans. Acta Pharmacol Scand 13:429–432

Mizuno Y, Sone N, Saitoh T (1986) Dopaminergic neurotoxins, MPTP and MPP$^+$, inhibit mitochondrial NADH-ubiquinone oxidoreductase activity. Proc Jap Acad (Ser B) 62:261–263

Mizuno Y, Sone N, Saitoh T (1987a) Inhibition of mitochondrial NADH-ubiquinone oxidoreductase activity by 1-methyl-4-phenylpyridinium ion. Biochem Biophys Res Commun 143:294–299

Mizuno Y, Sone N, Saitoh T (1987b) Effects of 1-methyl-4-phenyl-1,2,3,6-tetrahydropyridine and 1-methyl-4-phenylpyridinium ion on activities of the enzymes in the electron transport system in mouse brain. J Neurochem 48:1787–1793

Mizuno Y, Sone N, Saitoh T (1987c) Inhibition of mitochondrial alpha-ketoglutarate dehydrogenase by 1-methyl-4-phenylpyridinium ion. Biochem Biophys Res Commun 143:971–976

Mizuno Y, Sone N, Suzuki K, Saitoh T (1988) Studies on the toxicity of 1-methyl-4-phenylpyridinium ion (MPP$^+$) against mitochondria of mouse brain. J Neurol Sci 86:97–110

Mizuno Y, Ohta S, Tanaka M, Takamiya S, Suzuki K, Sato T, Oya H, Ozawa T, Kagawa Y (1989) Deficiencies in Complex I subunits of the respiratory chain in Parkinson's disease. Biochem Biophys Res Commun 163:1450–1455

Mizuno Y, Suzuki K, Ohta S (1990) Postmortem changes in mitochondrial respiratory enzymes in brain and a preliminary observation in Parkinson's disease. J Neurol Sci 96:49–57

Moraes CT, DiMauro S, Zeviani M, Lombes A, Shanske S, Miranda AF, Nakase H, Bonilla E, Werneck LC, Servidei S, Nonaka I, Koga Y, Spiro AJ, Brownell KW, Schmidt B, Schotland DL, Zupanc M, DeVivo DC, Schon EA, Rowland LP (1989) Mitochondrial DNA deletions in progressive external ophthalmoplegia and Kearns-Sayre syndrome. N Engl J Med 320:1293–1299

Nicklas WJ, Vyas I, Heikkila RE (1985) Inhibition of NADH-linked oxidation in brain mitochondria by 1-methyl-4-phenyl-1,2,3,6-tetrahydropyridine. Life Sci 36:2503–2508

Ozawa T, Yoneda M, Tanaka M, Ohno K, Sato W, Suzuki H, Nishikimi M, Yamamoto M, Nonaka K, Horai S (1988) Maternal inheritance of deleted DNA in a family with mitochondrial myopathy. Biochem Biophys Res Commun 154:1240–1247

Ozawa T, Tanaka M, Ikebe S, Ohno K, Kondo T, Mizuno Y (1990) Quantitative determination of deleted mitochondrial DNA relative to normal DNA in parkinsonian striatum by a kinetic PCR analysis. Biochem Biophys Res Commun 172:483–489

Parker WD Jr, Boyson SJ, Parks JK (1990) Abnormalities of the electron transport chain in idiopathic Parkinson's disease. Ann Neurol 26:719–723

Ramsay RR, Salach JI, Singer TP (1986) Uptake of neurotoxin 1-methyl-4-phenylpyridine (MPP$^+$) by mitochondria and its relation to the inhibition of the mitochondrial oxidation of NAD-linked substrates by MPP$^+$. Biochem Biophys Res Commun 134:743–748

Ramsay RR, Singer TP (1986) Energy-dependent uptake of N-methyl-4-phenylpyridinium, the neurotoxic metabolite of 1-methyl-4-phenyl-1,2,3,6-tetrahydropyridine, by mitochondria. J Biol Chem 261:7585–7587

Richter C, Park JW, Ames BN (1988) Normal oxidative damage to mitochondrial and nuclear DNA is extensive. Proc Natl Acad Sci USA 85:6465–6467

Schapira AHV, Cooper JM, Dexter D, Jenner P, Clark JB, Marsden CD (1989) Mitochondrial complex I deficiency in Parkinson's disease. Lancet i:1269

Schapira AHV, Cooper JM, Dexter D, Clark JB, Jenner P, Marsden CD (1990) Mito-
    chondrial complex I deficiency in Parkinson's disease. J Neurochem 54:823–827
Schofner JM, Lott MT, Lezza AMS, Seibel P, Ballinger SW, Wallace DC (1990)
    Myoclonic epilepsy and ragged-red fiber disease (MERRF) is associated with a
    mitochondrial DNA tRNA^{Lys} mutation. Cell 61:931–937
Seyama K, Suzuki K, Mizuno Y, Yoshida M, Tanaka M, Ozawa T (1989) Mitochondrial
    encephalomyopathy with lactic acidosis and stroke-like episodes with special refer-
    ence to the mechanism of cerebral manifestations. Acta Neurol Scand 80:561–568
Tanaka M, Sato W, Ohno L, Yamamoto T, Ozawa T (1989) Direct sequencing of
    deleted mitochondrial DNA in myopathic patients. Biochem Biophys Res Commun
    164:156–163

Correspondence: Prof. Y. Mizuno, MD, Department of Neurology, Juntendo Univer-
    sity School of Medicine, 2-1-1 Hongo, Bunkyo, Tokyo 113, Japan.

Tetrahydroisoquinoline and its derivatives: the occurrence and the metabolism in the brain, and the effects on catecholamine metabolism

M. Naoi[1], T. Niwa[2], M. Yoshida[3], and T. Nagatsu[4]

[1] Department of Bioscience, Nagoya Institute of Technology,
[2] Department of Internal Medicine, Nagoya University Branch Hospital, Nagoya,
[3] Department of Neurology, Jichi Medical School, Tochigi, and
[4] Department of Biochemistry, Nagoya University School of Medicine, Nagoya, Japan

Summary

1,2,3,4-Tetrahydroisoquinoline (TIQ) was identified in normal and parkinsonian human brains by gas chromatography-mass spectrometry (GC-MS). In human brain, N-methyl TIQ (NMTIQ) was formed by a N-methyltransferase in the cytosol fraction. The in vivo synthesis of NMTIQ was also confirmed in the brains of marmosets systematically administrated with TIQ. TIQ itself was not oxidized by monoamine oxidase, but NMTIQ was oxidized by both types A and B monoamine oxidase. The oxidative product, N-methylisoquinolinium ion (NMIQ$^+$), inhibited enzymes participating in catecholamine metabolism, such as tyrosine hydroxylase, DOPA decarboxylase and monoamine oxidase. The uptake of NMIQ$^+$ was mediated by the dopamine transport system. These data support our view that TIQ and its derivatives may be candidates of dopaminergic neurotoxin similar to 1-methyl-4-phenyl-1,2,3,6-tetrahydropyridine (MPTP).

Introduction

Characteristic pathological changes in Parkinson's disease are the severe loss of dopaminergic neurons and the decrease in dopamine (DA) contents in the striatum (Ehringer and Hornykiewicz, 1960). The reduction of tyrosine hydroxylase [L-tyrosine, tetrahydropteridine: oxygen oxidoreductase (3-hydroxylating) EC 1.14.16.2, TH], the rate-limiting enzyme in the biosynthesis of dopamine, and of tetrahydrobiopterin, a cofactor of TH were recognized in parkinsonian brains (Nagatsu et al., 1981). Since the discovery of a selective dopaminergic neurotoxin, 1-methyl-4-phenyl-1,2,3,6-tetrahydropyridine (MPTP) (Langston et al., 1983), exogenous or endogenous compounds have been sought as parkinsonism-inducing agents. From the data obtained by the use of an MPTP-induced parkinsonian model, several characteristics have been proposed for a toxin candidate to elicit degeneration of dopaminergic neurons in the nigro-striatal system. A compound is taken up into the human body and transported through the blood-brain barrier, or it is synthesized in vivo in the brain. It is metabolized in the brain to a more toxic compound and transported into dopaminergic neurons by the DA transport system. It is accumulated in a subcellular compartment, such as mitochondria, or in conjugation with neuromelanin, and induces cytotoxicity. To screen such dopaminergic neurotoxins, compounds whose structures are similar to MPTP or its oxidized product, 1-methyl-4-phenylpyridinium ion (MPP$^+$), were examined for their effects on catecholamine metabolism. By culture in the presence of MPP$^+$, the activities of the enzymes related to catecholamine metabolism were markedly reduced in clonal rat pheochromocytoma PC12h cells (Naoi et al., 1988). The concentrations of MPP$^+$ required to affect the enzyme activity were much lower than the MPP$^+$ concentration required for cell death. This indicates that the dopaminergic neurotoxicity of an agent may be screened by its effect on the catecholamine metabolism. Based on this idea, the N-methylisoquinolinium ion (NMIQ$^+$) was found to inhibit TH activity in rat striatal slices (Hirata et al., 1986). The results suggest that the compounds structurally related to NMIQ$^+$, such as 1,2,3,4-tetrahydroisoquinoline (TIQ) and its N-methylated compound, may be neurotoxins similar to MPTP.

This paper describes the occurrence of TIQ in the human brain, its metabolism into N-methyl-TIQ (NMTIQ) and further oxidation into NMIQ$^+$, and the effects on catecholamine metabolism. As shown in Fig. 1, NMTIQ and NMIQ$^+$ have chemical structures very similar to those of MPTP and MPP$^+$.

Occurrence of TIQ in human brain

Normal human brains and a brain from a patient with Parkinson's disease were analyzed by the use of gas chromatography-mass spectrography (GC-MS).

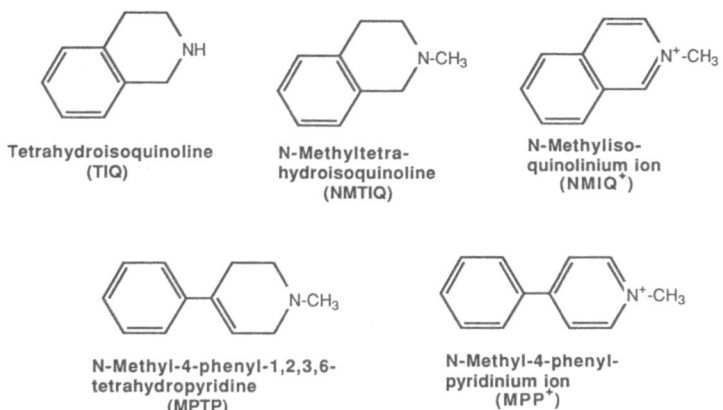

Fig. 1. Chemical structures of TIQ, NMTIQ, NMIQ$^+$, MPTP, and MPP$^+$

The brain samples were extracted, derivatized and analyzed as described by Niwa et al. (1987). Figure 2 shows mass chromatograms of the heptafluorobutyric (HFB) derivatives of TIQ and 2-methyl-1,2,3,4-tetrahydroquinoline (2-Me-TQ) (a), and the HFB-derivatized extracts from the frontal cortex of a patient with Parkinson's disease (b) and from a control (c). The identification of these compounds in the extract from the human brains was carried out by the use of electron-impact ionization (EI) mass spectra, as shown in Fig. 3, and by comparison of the retention times on GC, and the two peaks in the brain extract

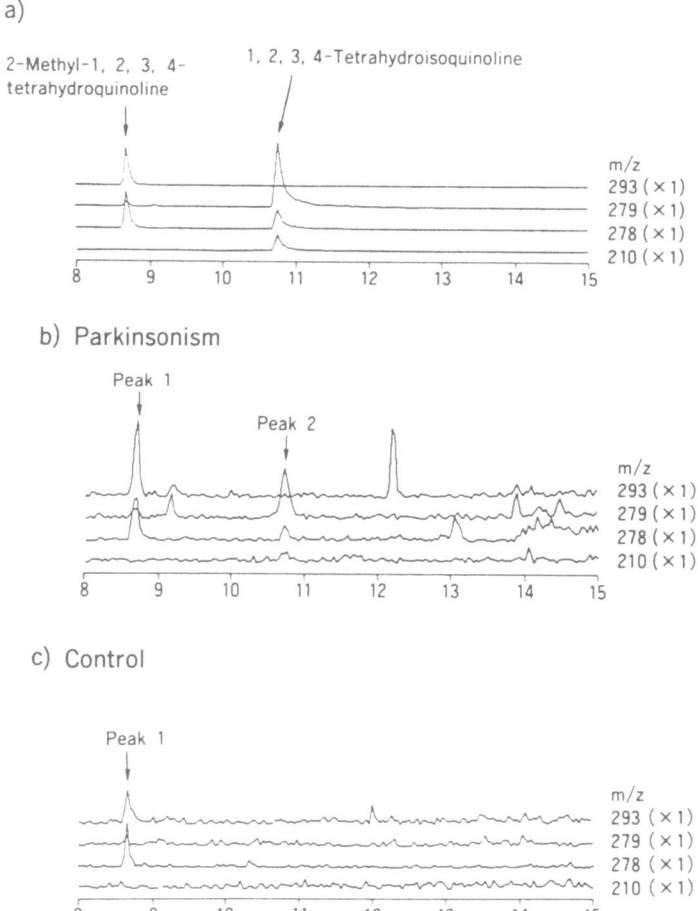

Fig. 2. Mass chromatograms of the HFB derivatives of 2-methyl-1,2,3,4-tetrahydroquinoline and 1,2,3,4-tetrahydroisoquinoline (a), the HFB-derivatized extracts from parkinsonian brain (b), and from control brain (c)

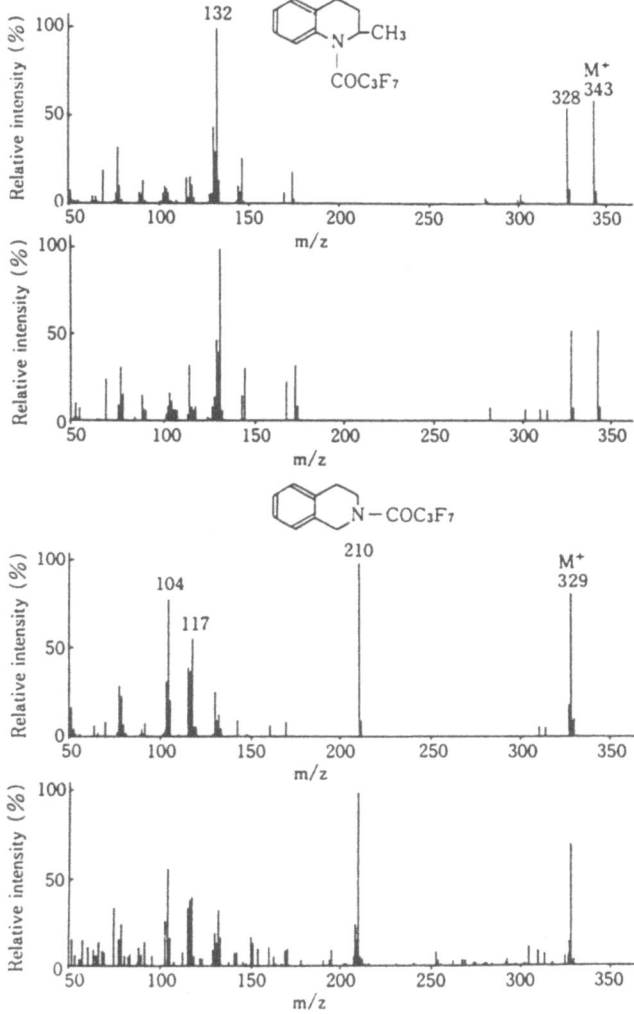

Fig. 3. EI mass spectra of the HFB derivative of 2-Me-TQ (a), and peak 1 in Fig. 2 (b), and that of TIQ (c) and peak 2 in Fig. 2 (d)

were identified to be 2-Me-TQ and TIQ. By selective ion monitoring, the concentration of TIQ in the frontal cortex of the patient with Parkinson's disease was increased compared with that in the control brain: approximately 10 ng/g and 1 ng/g, respectively. The presence of TIQ in parkinsonian brain as an endogenous amine was further confirmed by GC-MS after derivatization as a trimethylsilylated compound under milder conditions (Niwa et al., 1989a). The presence of TIQ in human brains was reported also by Ohta et al. (1987) and in rat brains (Kohno et al., 1986).

TIQ in the human brain may be derived from foods, taken up into the human body, and transported to the brain. The occurrence of TIQ in foods was confirmed by GC-MS (Niwa et al., 1989b). TIQ concentrations were found to be especially high in cheese, milk, the white of boiled egg, and banana: 5.2, 3.3, 2.2, and 2.2 ng/g, respectively. TIQ was transported into the brain through the blood-brain barrier. In the brains of monkeys systematically administered TIQ (Nagatsu and Yoshida, 1988), TIQ contents in the brains increased to 201 and 149 $\mu g/g$ wet weight from 0.15 and 0.19 $\mu g/g$ wet weight of control brain (Niwa et al., 1988). Thus, food-derived TIQ may be transported into the human brain and accumulated. Another possible origin of TIQ in the brain is its in vivo synthesis from β-phenylethylamine and formaldehyde by a Pictet-Spengler reaction (Zarranz de Ysern and Ordonez, 1981). By the Pictet-Spengler reaction of a β-arylethylamine with a carbonyl group, the biosynthesis of salsolinol (1-methyl-6,7-dihydroxy-1,2,3,4-tetrahydroisoquinoline) was reported (Cohen and Collins, 1970). This reaction occurs at a significant rate in vivo (Collins, 1980). However, the in situ synthesis of TIQ in human brain has not been confirmed, and the conclusion on the origin of TIQ in the human brain awaits further experiments.

TIQ was systematically administrated to marmosets and parkinsonism was produced (Nagatsu and Yoshida, 1988). The activity of TH and the total contents of DA and biopterins were reduced significantly. The selective sensitivity of dopaminergic neuron to TIQ was demonstrated by immunohistochemical studies in C57BL/6j mouse brains systematically administrated TIQ (Ogawa et al., 1989). In the substantia nigra pars compacta and ventral tegmental area of TIQ-treated mice, the number of TH-like immunoreactive neurons decreased significantly. However, the amounts of TIQ used were rather high: 50 mg/kg per day for 11 days and 70 days in the experiments with marmosets and mice, respectively. Chronic administration of TIQ (20 mg/kg per day, subcutaneously for up to 104 days) to squirrel monkeys produced motor symptoms similar to parkinsonism even after 7 days of discontinuation of TIQ, and the symptoms were alleviated remarkably by L-DOPA treatment (Yoshida et al., 1990). Cell death of the dopaminergic neurons did not occur in mice, which may be due to the age of the animals, or to less potent and less specific neurotoxicity of TIQ to dopaminergic neurons. From this point of view, the possible neurotoxicity of other derivatives of TIQ was studied.

Metabolism of TIQ into NMTIQ and NMIQ$^+$ in human brain

Because of the occurrence of TIQ in the human brain, in situ biosynthesis of NMIQ$^+$, a potent dopaminergic neurotoxin, from TIQ has been supposed. TIQ was neither bound to nor oxidized by either type A or B monoamine oxidase [monoamine: oxygen oxidoreductase (deaminating), EC 1.4.3.4., MAO] (Naoi and Nagatsu, 1987), so two steps of biosynthesis have been proposed: N-methylation of TIQ and its oxidation into NMIQ$^+$.

Table 1. Kinetic data on N-methyltransferase activity in human brain homogenate

In terms of S-adenosyl-L-methionine	
K_m	$5.11 \pm 1.69\,\mu M$
V_{max}	7.31 ± 0.21 pmol/min/mg protein
In terms od TIQ	
K_m	$20.9 \pm 5.5\,\mu M$
V_{max}	7.98 ± 1.21 pmol/min/mg protein

The values represent the mean and SD

Brain homogenate was incubated with TIQ in the presence of a methyl donor, S-adenosyl-L-methionine (SAM) and the reaction products were analyzed using high-performance liquid chromatography (HPLC) and electrochemical detection, as reported previously (Naoi et al., 1989b). The kinetic data on N-methylation of TIQ were summarized in Table 1. The major part of N-methyl-transferase activity is localized in the soluble and cytosol fraction of the human brain.

NMTIQ was incubated with type A and B MAO samples prepared from human brain synaptosomal mitochondria, and its oxidative product, $NMIQ^+$,

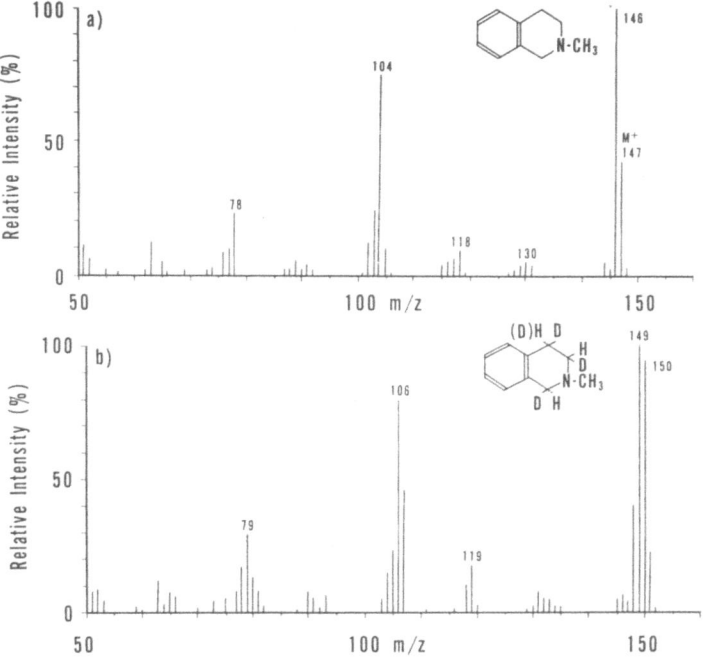

Fig. 4. EI mass spectra of NMTIQ (a) and a mixture of d_3-NMTIQ and d_4-NMTIQ (b)

was quantitatively analyzed by HPLC with fluorescent detection (Naoi et al., 1989a). The values of the maximal velocity, V_{max}, and the Michaelis constant, K_m, of the oxidation of TIQ are $571 \pm 25\,\mu M$ and 0.29 ± 0.06 pmol/min/mg protein, respectively, for type A; $463 \pm 43\,\mu M$ and 0.16 ± 0.03 pmol/min/mg protein, respectively, for type B. NMTIQ inhibited the oxidation of an amine substrate, kynuramine, by type A and B MAO, and the inhibition was competitive, indicating that TIQ and amine substrates share the same binding site of MAO.

The in vivo synthesis of NMTIQ from TIQ was confirmed by the analysis of the brains of TIQ-induced parkinsonian marmosets. By quantitative analysis using GC-MS (Niwa et al., 1990), NMTIQ was detected in the brains of monkeys treated with TIQ. Figure 4 shows EI mass spectra of authentic NMTIQ (a) and a mixture of d_3-NMTIQ and d_4-NMTIQ (b). The EI mass spectrum of NMTIQ shows characteristic ions at m/z 147 (M^+) and 146. Figure 5 shows SIM chromatograms of authentic NMTIQ, a mixture of d_3-NMTIQ and d_4-NMTIQ (a), and the extract from the brain of a TIQ-treated monkey (b) using an OV-101 column. NMTIQ was detected in the extract from the brain of a TIQ-treated monkey, since the peak showed an identical retention time (4.4 min) and height ratio (m/z 147/146) as authentic NMTIQ. A mixture of d_3-NMTIQ and d_4-NMTIQ was included as an internal standard in the extract of TIQ-treated monkeys to quantitate the NMTIQ level. The NMTIQ concentrations in the brains were 1.7 and $2.2\,\mu g/g$ wet weight, which correspond to about 1% of the amounts of TIQ detected in the same brain.

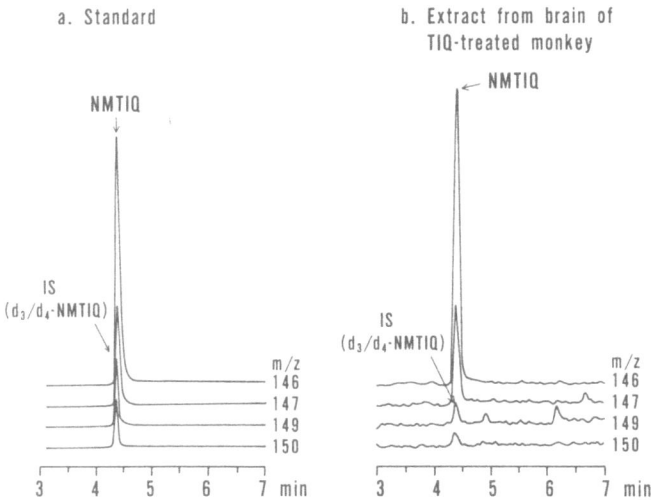

Fig. 5. SIM chromatograms of authentic NMTIQ, a mixture od d_3-NMTIQ and d_4-NMTIQ (a), and extract from the brain of a TIQ-treated monkey (b) using an OV-101 capillary column

Effect of NMIQ$^+$ on catecholamine metabolism

The effect of NMIQ$^+$ on tyrosine hydroxylation was demonstrated using rat striatal slices (Hirata et al., 1986; Nagatsu and Hirata, 1987). NMIQ$^+$ inhibited the DOPA formation from tyrosine in a dose-dependent manner. NMIQ$^+$ at 10 μM reduced tyrosine hydroxylation to 60% of the control and at 100 μM to 25%. The effects of NMIQ$^+$ on activities of enzymes related to the catecholamine metabolism were studied using a clonal rat pheochromocytoma cell line, PC12h (Naoi et al., 1989c). Table 2 summarizes the kinetic data of PC12h cells and K$_i$ values of NMIQ$^+$. NMIQ$^+$ inhibited TH activity in the intact cells in a dose-dependent way, and the IC$_{50}$ was calculated to be 75 μM. However, the mechanism of the inhibition by TIQ has not been well elucidated. One possible inhibition site of tyrosine hydroxylation may be the inhibition of its phosphorylation, as in the case of MPP$^+$ (Kiuchi et al., 1989). NMIQ$^+$ inhibited the activity of DOPA decarboxylase [aromatic L-amino acid carboxylyase, EC 4.1.1.28, DDC] in competition with a co-factor, pyridoxal-5-phosphate, and it inhibited type A MAO in PC12h cells; the inhibition was competitive with a substrate, kynuramine. Type A MAO in human brain synaptosomal mitochondria was more sensitive to NMIQ$^+$ than type B; the K$_i$ value of NMIQ$^+$ for type A and B was 40.6 ± 4.9 and $284.6 \pm 9.5 \mu$M, respectively (Naoi et al., 1987). Since type A MAO oxidizes monoamines

Table 2. Kinetic data of TH, DDC and MAO in PC12h cells and effect of NMIQ$^+$

	K$_m$ (μM)	V$_{max}$ (pmol/min/mg protein)
Tyrosine hydroxylase		
Toward L-tyrosine	19.5 ± 6.9	264 ± 68
Toward (6R)-4-erythro-5,6,7,8-tetrahydrobiopterin		
	361 ± 58	510 ± 38
	87 ± 6.4	267 ± 74
IC$_{50}$ of NMIQ$^+$*	75 ± 50 μM	
DOPA decarboxylase		
Toward L-DOPA, using 10 μM pyridoxal-5-phosphate		
	21.4 ± 1.6	1310 ± 10
Toward pyridoxal-5-phosphate, using 1 mM L-DOPA		
	0.62 ± 0.10	436 ± 58
K$_i$ of NMIQ$^+$ in terms of pyridoxal-5-phosphate		
	89.9 ± 7.5 μM	
Monoamine oxidase		
Toward kynuramine	17.8 ± 0.9	869 ± 30
K$_i$ of NMIQ$^+$	20.0 ± 6.5 μM	

*IC$_{50}$ is the concentration of NMIQ$^+$ required to yield 50% inhibition of TH activity in the intact cells.

Each value represents the mean and SD of triplicate measurements of two experiments, using eight different concentrations

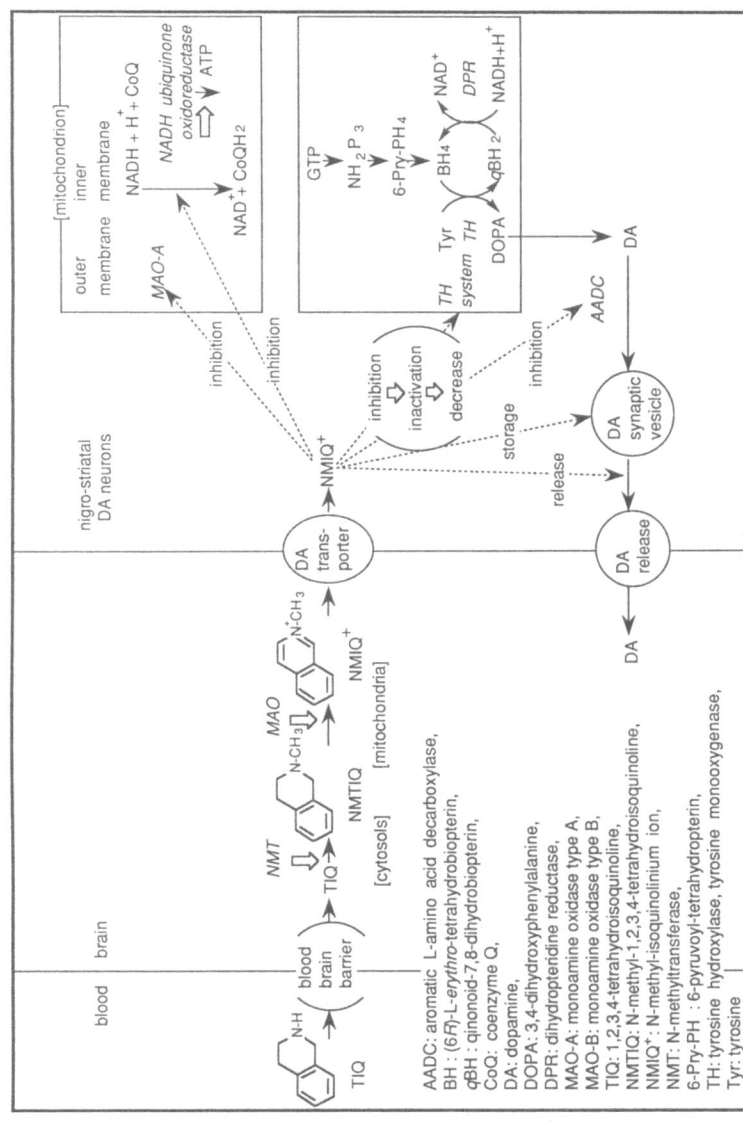

Fig. 6. The metabolism and effects on catecholamine metabolism of TIQ and its metabolites

AADC: aromatic L-amino acid decarboxylase,
BH: (6R)-L-erythro-tetrahydrobiopterin,
qBH: qinonoid-7,8-dihydrobiopterin,
CoQ: coenzyme Q,
DA: dopamine,
DOPA: 3,4-dihydroxyphenylalanine,
DPR: dihydropteridine reductase,
MAO-A: monoamine oxidase type A,
MAO-B: monoamine oxidase type B,
TIQ: 1,2,3,4-tetrahydroisoquinoline,
NMTIQ: N-methyl-1,2,3,4-tetrahydroisoquinoline,
NMIQ⁺: N-methyl-isoquinolinium ion,
NMT: N-methyltransferase,
6-Pry-PH : 6-pyruvoyl-tetrahydropterin,
TH: tyrosine hydroxylase, tyrosine monooxygenase,
Tyr: tyrosine

functioning as neurotransmitters such as serotonin and adrenaline, the inhibition may perturb the levels of these monoamines in the brain.

By culturing for six days in the presence of $NMIQ^+$ in the culture medim, the TH activity was reduced at concentrations higher than $10\,\mu M$, and the activity was about 50% of the control. However, the activity of DDC and MAO was reduced only in the presence of 1 mM $NMIQ^+$. $NMIQ^+$ is taken up into rat striatal slices (Hirata et al., 1990) and the uptake is mediated by the DA transport system, which was proven by specific inhibition of $NMIQ^+$ uptake with DA itself and mazindol, an inhibitor of DA uptake (Naoi et al., 1989c).

Can TIQ be a naturally occurring dopaminergic neurotoxin?

The data presented here are summarized in Fig. 6, and they suggest that TIQ may be metabolized into $NMIQ^+$ in the brain, transported in dopaminergic neurons, and may inhibit tyrosine hydroxylation and other steps of catecholamine metabolism. The inhibition of DDC and MAO activities by $NMIQ^+$ is weaker than by MPP^+ (Naoi et al., 1988). However, the in vivo effect of $NMIQ^+$ on TH activity in the cells, shown by 6 days of culturing in the presence of $NMIQ^+$, was almost the same as MPP^+: IC_{50} was about $10\,\mu M$ for both of the neurotoxins. The cytotoxicity of MPTP and MPP^+ is selective to dopaminergic neurons in substantia nigra zona compacta (Burns et al., 1983), and the mechanism of its specificity is considered to be ascribed to the specific uptake into dopaminergic neurons (Javitch et al., 1985) and to accumulation in neuromelanin. The mechanism of cell death may be neither selective nor specific to dopaminergic neurons, but a rather general mechanism may be involved, such as lipid peroxidation or energy crisis, as in other neurodegenerative process. In fact, TIQ inhibited the activity of Complex I and the synthesis of ATP (Suzuki et al., 1988). Although TIQ and especially $NMIQ^+$ proved to be potent inhibitors of the catecholamine metabolism, selective accumulation in intracellular compartments or more toxicity may be required to elicit the cytotoxicity. Other TIQ derivatives with more affinity and cytotoxicity to dopaminergic neurons should be determined by systematic screening of compounds.

Acknowledgment

We would like to thank Dr. S. Matsuura, Department of Chemistry, College of General Education, Nagoya University, for his kind supply of NMTIQ and $NMIQ^+$.

References

Burns RS, Chiueh CC, Markey SP, Ebert MH, Jacobowitz DM, Kopin IJ (1983) A primate model of parkinsonism: selective destruction of dopaminergic neurons in the pars compacta of the substantia nigra by N-methyl-4-phenyl-1,2,3,6-tetrahydropyridine. Proc Natl Acad Sci USA 80:4546−4550

Cohen G, Collins MA (1970) Alkaloids from catecholamines in adrenal tissue: possible role in alcoholism. Science 167:1749−1751

Collins MA (1980) Neuroamine condensations in human subjects. Adv Exp Med Biol 126:87−102

D'Amato RJ, Lipman ZP, Snyder SH (1986) Selectivity of the parkinsonian neurotoxin MPTP: toxic metabolite MPP$^+$ binds to neuromelanin. Science 231:987−989

Ehringer H, Hornykiewicz O (1960) Verteilung von Noradrenalin und Dopamin (3-Hydroxytyramin) im Gehirn des Menschen und ihr Verhalten bei Erkrankungen des extrapyramidalen System. Klin Wschr 24:1236−1239

Hirata Y, Sugimura H, Takei H, Nagatsu T (1986) The effects of pyridinium salts, structurally related compounds of 1-methyl-4-phenylpyridinium ion (MPP$^+$), on tyrosine hydroxylation in rat striatal tissue slices. Brain Res 397:341−344

Hirata Y, Minami M, Naoi M, Nagatsu T (1990) Studies on the uptake of N-methylisoquinolinium ion into rat striatal slices using high-performance liquid chromatography with fluorometric detection. J Chromatogr 503:189−195

Javitch JA, D'Amato RJ, Strittmatter SM, Snyder SH (1985) Parkinsonism-inducing neurotoxin, N-methyl-4-phenyl-1,2,3,6-tetrahydropyridine: uptake of the metabolite N-methyl-4-phenylpyridine by dopamine neurons explains selective toxicity. Proc Natl Acad Sci USA 82:2173−2177

Kiuchi K, Hagihara M, Hidaka H, Nagatsu T (1988) Effect of 1-methyl-4-phenylpyridinium ion on tyrosine hydroxylase in rat pheochromocytoma PC12h cells. Neurosci Lett 89:209−215

Kohno M, Ohta S, Hirobe M (1986) Tetrahydroisoquinoline and 1-methyl-tetrahydroisoquinoline as novel endogenous amines in rat brain. Biochem Biophys Res Commun 140:448−454

Langston JW, Ballard P, Tetrud JW, Irwin I (1983) Chronic parkinsonism in humans due to a product of meperidine-analog synthesis. Science 219:979−980

Nagatsu T, Hirata Y (1987) Inhibition of the tyrosine hydroxylase system by MPTP, 1-methyl-4-phenylpyridinium ion (MPP$^+$) and the structurally related componds in vitro and in vivo. Eur Neurol 26 [Suppl 1]:11−15

Nagatsu T, Yamaguchi T, Kato T, Sugimoto T, Matsuura S, Akino M, Nagatsu I, Iizuka R, Narabayashi H (1981) Biopterin in human brain and urine from controls and parkinsonian patients: application of a new radioimmunoassay. Clin Chim Acta 109:305−311

Nagatsu T, Yoshida M (1988) An endogenous substance of the brain, tetrahydroisoquinoline, produces parkinsonism in primates with decreased dopamine, tyrosine hydroxylase and biopterin in the nigrostriatal regions. Neurosci Lett 87:178−182

Naoi M, Nagatsu T (1987) Quinoline and quinaldine as naturally occurring inhibitors specific for type A monoamine oxidase. Life Sci 40:1075−1082

Naoi M, Hirata Y, Nagatsu T (1987) Inhibition of monoamine oxidase by N-methylisoquinolinium ion. J Neurochem 48:709−712

Naoi M, Takahashi T, Nagatsu T (1988) Effect of 1-methyl-4-phenylpyridinium ion (MPP$^+$) on catecholamine levels and activity of related enzymes in clonal rat pheochromocytoma PC12h cells. Life Sci 43:1485−1491

Naoi M, Matsuura S, Parvez H, Takahashi T, Hirata Y, Minami M, Nagatsu T (1989a) Oxidation of N-methyl-1,2,3,4-tetrahydroisoquinoline into the N-methylisoquinolinium ion by monoamine oxidase. J Neurochem 52:653−655

Naoi M, Matsuura S, Takahashi T, Nagatsu T (1989b) A N-methyltransferase in human brain catalyses N-methylation of 1,2,3,4-tetrahydroisoquinoline into N-methyl-1,2,3,4-tetrahydroisoquinoline, a precursor of a dopaminergic neurotoxin, N-methylisoquinolinium ion. Biochem Biophys Res Commun 161:1213–1219

Naoi M, Takahashi T, Parvez H, Kabeya R, Taguchi E, Yamaguchi K, Hirata Y, Minami M, Nagatsu T (1989c) N-Methylisoquinolinium ion as an inhibitor of tyrosine hydroxylase, aromatic L-amino acid decarboxylase and monoamine oxidase. Neurochem Int 15:315–320

Niwa T, Takeda N, Kaneda N, Hashizume Y, Nagatsu T (1987) Presence of tetrahydroisoquinoline and 2-methyl-tetrahydroquinoline in parkinsonian and normal human brains. Biochem Biophys Res Commun 144:1084–1089

Niwa T, Takeda N, Tatematsu A, Matsuura S, Yoshida M, Nagatsu T (1988) Migration of tetrahydroisoquinoline, a possible parkinsonian neurotoxin, into monkey brain from blood as proved by gas chromatography-mass spectrometry. J Chromatogr 452:85–91

Niwa T, Takeda N, Sasaoka T, Kaneda N, Hashizume Y, Yoshizumi H, Tatematsu A, Nagatsu T (1989a) Detection of tetrahydroisoquinoline in parkinsonian brain as an endogenous amine by gas chromatography-mass spectrometry. J Chromatogr 491:397–403

Niwa T, Yoshizumi H, Tatematsu A, Matsuura S, Nagatsu T (1989b) Presence of tetrahydroisoquinoline, a parkinsonism-related compound, in foods. J Chromatogr 493:347–352

Niwa T, Yoshizumi H, Tatematsu A, Matsuura S, Yoshida M, Kawachi M, Naoi M, Nagatsu T (1990) Endogenous synthesis of N-methyl-1,2,3,4-tetrahydroisoquinoline, a precursor of N-methylisoquinolinium ion, in the brains of primates with parkinsonism after systemic administration of 1,2,3,4-tetrahydroisoquinoline. J Chromatogr 533:145–151

Ohta S, Kohno M, Makino Y, Tachikawa O, Hirobe M (1987) Tetrahydroisoquinoline and 1-methyl-tetrahydroisoquinoline are present in the human brain: relation to Parkinson's disease. Biomed Res 8:453–456

Ogawa M, Araki M, Nagatsu I, Nagatsu T, Yoshida M (1989) The effect of 1,2,3,4-tetrahydroisoquinoline (TIQ) on mesencephalic dopaminergic neurons in C57BL6J mice: Immunohistochemical studies-tyrosine hydroxylase. Biogenic Amines 6:427–436

Suzuki K, Mizuno Y, Yoshida M (1988) Inhibition of mitochondrial NADH-ubiquinone oxidoreductase activity and ATP synthesis by tetrahydroisoquinoline. Neurosci Lett 86:105–108

Yoshida M, Niwa T, Nagatsu T (1990) Parkinsonism in monkeys produced by chronic administration of an endogenous substance of the brain, tetrahydroisoquinoline: the behavioral and biochemical changes. Neurosci Lett 119:109–113

Zarranz de Ysern ME, Ordnez l (1981) Tetahydroisoquinolines: a review. Prog Neuropsychopharmacol 5:343–355

Correspondence: Dr. M. Naoi, Department of Bioscience, Nagoya Institute of Technology, Showa-ku, Nagoya 466, Japan

Can tetrahydroisoquinoline (TIQ) or N-methyl-TIQ produce parkinsonism?

M. Yoshida[1], M. Ogawa[1], and T. Nagatsu[2]

[1] Department of Neurology, Jichi Medical School, Tochigiken, and
[2] Department of Biochemistry, Nagoya University, Nagoya, Japan

Summary

We have examined the effect of chronic administration of a probable endogenous dopaminergic neurotoxin, tetrahydroisoquinoline (TIQ, 20 mg/kg per day, s.c. for up to 104 days), on squirrel monkeys. Chronically administered TIQ produced motor symptoms similar to parkinsonism in squirrel monkeys even after 7 days' discontinuation of TIQ and the symptoms were alleviated remarkably by levodopa treatment. Biochemical analysis of the brains of TIQ-treated monkeys revealed significant decrease in dopamine (DA) and total biopterin (BP) concentrations, and tyrosine hydroxylase (TH) activity in the substantia nigra. Morphological changes of the substantia nigra of the mice were also analysed. The effect of long-term administration of N-methyl-TIQ on aged monkeys was described.

Introduction

The discovery (Langston et al., 1983) that 1-methyl-4-phenyl-1,2,3,6-tetrahydropyridine (MPTP), a side product of meperidine, produced symptoms quite similar to those of Parkinson's disease in humans promoted research toward elucidation of the etiology of Parkinson's disease. MPTP, however, can not be a causal substance to produce Parkinson's disease, since MPTP does not exist in the natural environment. Nagatsu and Hirata (1987) have extensively screened substances with a similar structure as MPTP and found from *in vitro* as well as *in vivo* experiments that tetrahydroisoquinoline (TIQ) or N-methyl TIQ (NMeTIQ), whose structures are shown in Fig. 1, might be candidates to produce parkinsonism.

Recently, we administered TIQ to monkeys and mice, and found that TIQ produced remarkable motor disturbances that were reversed by levodopa administration, biochemical changes in the substantia nigra of the monkeys, and neuropathological changes in the substantia nigra of the mice.

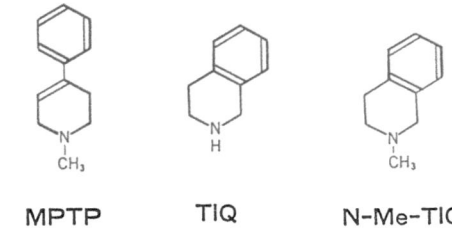

MPTP   TIQ   N-Me-TIQ

Fig. 1. Chemical structures of MPTP, TIQ and NMeTIQ

Presence of TIQ in the brain as well as in foods

Gas chromatography-mass spectrometry has demonstrated the presence of TIQ in the brain of rats as well as humans (Kohno et al., 1986; Ohta et al., 1987; Niwa et al., 1987). In rats the TIQ concentrations were about 6 ng/g wet weight of tissue in the spinal cord and also 2 ng/g in the brain, while those in the lung, intestine, and liver were about 1 ng/g or less. The concentrations of TIQ in the frontal cortex of the brains of the control subjects and the patients with Parkinson's disease were 0.86 ± 0.23 and 0.54 ± 0.20 ng/g, respectively, and the corresponding values for the caudate nucleus were 0.64 ± 0.24. Although there was a tendency that the TIQ concentration was lower in Parkinson's disease than that in the controls, no definite conclusion could be drawn because of the low sample numbers. Aside from its presence in the human body, TIQ has also been detected in cheese, wine, and cocoa (Makino et al., 1988). Furthermore, Niwa et al. (1989) recently found that TIQ was distributed widely in a variety of foods.

Behavioral changes associated with long-term administration of TIQ in monkeys

We have examined the effect of chronic administration of TIQ on squirrel monkeys (Yoshida et al., 1990). Eighteen squirrel monkeys with the age ranging from 4 to 5.5 years (mean 4.53 years, 13 males and 5 females) and weighing 450–700 g (mean 566.7 g) were studied. TIQ was administered to 13 monkeys (20 mg/kg per day, s.c.) for up to 104 days. Saline was injected subcutaneously once a day into 5 other monkeys used as controls. The behavioral changes were analyzed by using a conventional video-tape recorder, and videos were taken daily and evaluated according to the activities described in the legend to Table 1. Each item was evaluated by 5 grades from 0 to 4 such that the greater the motor dysfunction or severity of symptoms, the higher the value. Thus, a total score of 36 would indicate the worst possible condition. While all 5 saline-injected monkeys had a score of 0, the 13 TIQ-treated animals showed scores from 20 to 35 (mean $\pm$ S.D. 27.8 ± 4.2, n $=$ 13).

Table 1. Behavioral score of motor disability of saline- or tetrahydroisoquinoline (*TIQ*)-injected monkeys

Monkeys Injection	A Saline	B	C	D	E	F	G	H	I	J	K	L	M	N
									TIQ					
Days of injection	60	43	43	43	58	58	60	60	60	60	60	60	104	104
Days after the last injection*	1	1	1	1	1	1	1	(7)	(7)	(7)	(7)	(7)	(7)	(7)
Numbers	5	1	1	1	1	1	1	1	1	1	1	1	1	1
a	0	3	3	3.5	4	3	4	3	3.5	3.5	4	4	3	4
b	0	3	3	3.5	4	3	4	3	3.5	3.5	4	4	3	4
c	0	4	4	4	4	4	3	2	3	3.5	3	4	3	4
d	0	4	4	4	4	4	3	2	3	3.5	3	4	2	4
e	0	3	3	3	3	3	3	2	2	2.5	3	3	3	4
f	0	3	3	3	3	3	3	2	2	2.5	3	3	2	4
g	0	3	4	4	4	4	4	2	3	3	4	4	2	4
h	0	2	2	2	4	2	3	2	2	2	3	4	2	4
i	0	2	2	1	2	2	3	2	2	2	3	2	2	3
Total score	0/36	27/36	28/36	28/36	32/36	28/36	30/36	27/36 (20/36)	27.5/36 (24/36)	29/36 (26/36)	30/36 (30/36)	34/36 (32/36)	27/36 (22/36)	36/36 (35/36)

Eighteen squirrel monkeys were studied: *A* those injected with saline ($n = 5$); *B–N* those injected with TIQ ($n = 13$). Saline alone was injected subcutaneously once a day in 5 monkeys (A), and TIQ (20 mg/kg) in monkeys in B–N. One or 7 days after the last injection, behaviors were analyzed by using a video-tape recorder and evaluated according to the scoring procedure indicated: *a* climbing the cage wall a vertical shift (0, frequently; 2, rarely; 4, not observed); *b* moving on the floor a horizontal shift (0, promptly; 2, slowly; 4, not observed); *c* speed of limb-movement (0, fast; 2, slow; 4, no movement); *d* amount of limb-movement (0, natural and large; 2, small; 4, not observed); *e* speed of head-movement (0, natural and large; 2, small; 4, not observed); *f* amount of head-movement (0, natural and large; 2, small; 4, not observed); *g* grooming (0, frequent; 2, rare; 4, not observed); *b* lying down (0, not observed; 2, sometimes; 4, always); *i* tonus of limbs (0, normal; 2, slightly hyper; 4, hyper). In-between values, such as 1 and 3, were also scored if necessary. A total score of 36 would indicate the worst condition. *Time of analysis. The total scores 7 days after the last injection are indicated in parentheses in the bottom row of H–N

Effects of L-DOPA on TIQ-treated monkeys

Effects of L-DOPA were evaluated in some TIQ-treated monkeys listed in Table 1. As shown in Fig. 2A, L-DOPA (40 mg/kg) was orally administered to 3 monkeys (H, K and M in Table 1) 7 days after the final injection of TIQ given daily for 60 (H and K) and 104 days (M). Original motor disability scores improved gradually for 150 min after the levodopa administration. The motor disability scores returned to the original levels the next day.

In 2 monkeys (N and M, Fig. 2B,C) L-DOPA (20 mg) with benserazide (a DOPA decarboxylase inhibitor, 5.7 mg) was orally administered 8 days after

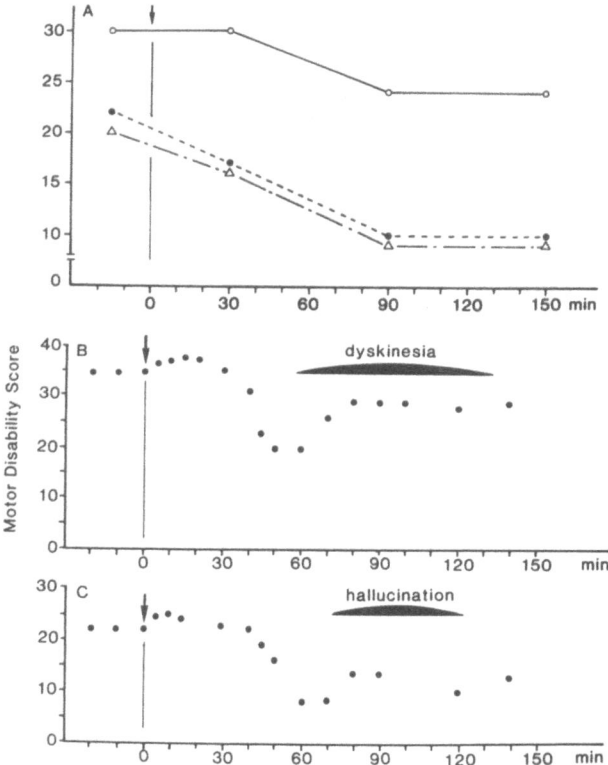

Fig. 2. Effect of L-DOPA on tetrahydroisoquinoline (TIQ)-treated monkeys. TIQ was given daily for 60 days (H, K) and 104 days (M, N). A L-DOPA (40 mg/kg) was orally administered (a downward arrow) to 3 TIQ-treated monkeys 7 days after the last injection of TIQ. Hollow circle, monkey K in Table 1. Filled circle, monkey M; triangle, monkey H. B L-DOPA (20 mg) and benserazide (5.7 mg) were orally administered 7 days after the last injection of TIQ in monkey N in Table 1. "Dyskinesia", 1 to 2 Hz dorsiflexion of the head. C same as in B, but in monkey M 8 days after the last injection of TIQ. "Hallucination", looking around restlessly as if hallucinating. Ordinate, motor disability scores evaluated as described in Table 1. A total score of 36 would indicate the worst possible condition

the final injection of TIQ given daily for 104 days. For the first 20 min after the administration both animals showed slight worsening of the motor activities which might have been due to increased hypokinesia. Thereafter, however, the symptoms improved remarkably for more than 150 min when observations were discontinued.

Biochemical studies in TIQ-treated monkeys

Biochemical studies were carried out on TIQ, dopamine (DA), total biopterin (BP) concentrations and tyrosine hydroxylase (TH) activity in the brain of controls and TIQ-treated parkinsonian monkeys. All animals were sacrificed under deep ketamine-induced anesthesia 1 or 7 days after the final administration of either saline or TIQ. Each brain was immediately removed, and kept at $-80°C$ until biochemical analyses were performed. TIQ, DA, total BP and TH activity were measured according to the methods described previously (Yoshida et al., 1990).

Biochemical data on TIQ, DA and total BP concentrations, and TH activity in the substantia nigra (SN) and caudate-putamen (C-P) are shown in Table 2. TIQ was identified in the brain of control squirrel monkeys by gas chromatography mass spectrometry. The endogenous levels of TIQ in the C-P of control monkeys were about $0.15 \mu g/g$, and increased up to $3-4 \mu g/g$ 1 day as well as 7 days following termination of TIQ administration for 43, 60 and 104 days. These data agree with our previous report that TIQ easily crosses the blood-brain barrier (Niwa et al., 1988). The metabolic clearance of TIQ in the monkey brain may be slow, since high TIQ concentrations were found 7 days after termination of TIQ administration. The reduction of DA and total BP concentrations, and TH activity in the SN was significant 1 day as well as 7 days following termination of TIQ administration that lasted for 43 and 60 days, respectively. The mean decreases in DA and total BP concentrations and TH activity 7 days after the last of the TIQ injections given for 60 days were 23%, 25%, and 53%, respectively. In contrast to the levels in the SN, the DA and total BP concentrations and TH activity in the C-P did not change significantly even after continuous injections for 60 days.

The present results are essentially similar to those given in our previous report on sub-acute experiments on marmosets (Nagatsu and Yoshida, 1988) with respect to the production of motor disturbances similar to parkinsonism accompanied by decreases in DA and BP concentrations and TH activity only in the SN.

Neuropathological changes of mesencephalic dopaminergic neurons of mice associated with long-term administration of TIQ

Ogawa et al. (1989) studied morphological changes in the brains of mice following chronic administration of TIQ. Male 8-week-old C57BL/6J mice

Table 2. Biochemical effects of tetrahydroisoquiroline (TIQ) in monkeys

	Days for injection	Day(s) after the last injection	Number of animals	TIQ (µg/g)	DA (pmol/mg protein)		BP (pmol/mg protein)		TH (pmol/min per mg protein)	
				CP	SN	C–P	SN	C–P	SN	C–P
Saline (control)	60	1	5	0.153 ±0.041	153 ±8	413 ±25	8.94 ±0.97	6.33 ±0.51	527 ±17	448 ±43
TIQ	43	1	3	3.90 ±0.92**	117 ±11*	493 ±14	7.43 ±0.43	7.31 ±0.79	288 ±27**	493 ±14
TIQ	58	1	2		129	178	5.02	3.80	332	248
					193	541	4.44	6.11	315	659
TIQ	60	1	1		113	601	4.73	4.32	207	492
TIQ	60	7	5	3.04 ±1.04**	116 ±6**	616 ±69	6.30 ±0.23*	7.63 ±0.93	248 ±28**	710 ±82
TIQ	104	7	2	2.70	100	321	7.58	6.68	246	405
				3.60	91	332	5.17	4.62	187	528

* $P < 0.02$, ** $P < 0.01$; significantly different from saline-treated controls. Each value is the mean ± S.D.

were used. TIQ was injected subcutaneously for 70 days at a dose of 50 mg/ kg/day. Twenty-four hours after the last injection of TIQ, the mice were anesthetized with sodium pentobarbital and their brains were perfused trans-cardially for 5 min with ice-chilled fixative containing 5% glutaraldehyde in 0.1 M phosphate buffer (pH 7.3). Following removal of the brain, immersion

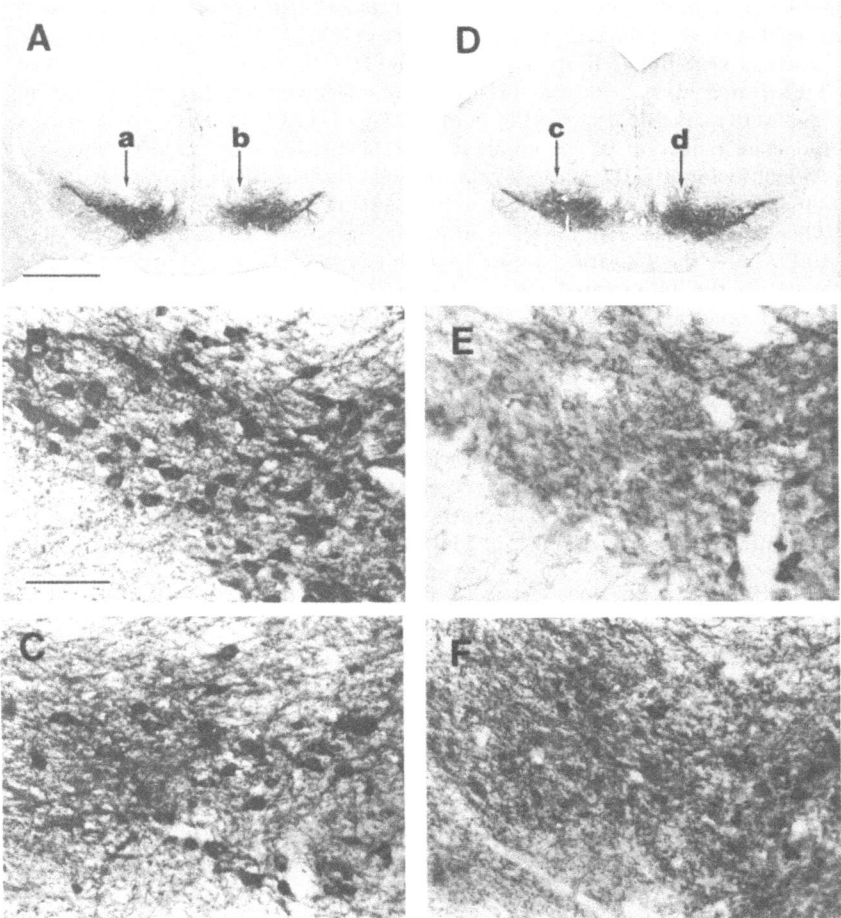

Fig. 3. Tyrosine hydroxylase (TH)-like immunohistochemistry of sections of the mesencephalon of a control (A,B,C) and a TIQ-treated mouse (D,E,F) specifically showing the SNc and VTA regions. (B) Magnification of A at the area indicated by arrow a. (C) Magnification of A at the area indicated by arrow b. (E) Magnification of D at the area indicated by arrow c. (F) Magnification of D at the area indicated by arrow d. Note the marked decrease in TH-like immunoreactivity of SNc cells (E). A moderate decrease is also seen in the cells of the VTA (F). A and D, × 12. Bar = 1000 μm in A,B,C,E and F; × 128. Bar = 100 μm in B

fixation was carried out for 24 h at 4°C in the same fixative. After washing of the tissue with 0.1 M phosphate buffer containing 15% sucrose, sections were cut at a 20-μm thickness with a freezing microtome. Immunostaining of free-floating sections was carried out by the peroxidase-anti-peroxidase (PAP) method. The primary antibody used in the present study was rabbit anti-tyrosine hydroxylase (diluted 2000 times). The specificity of this antibody was confirmed previously (Nagatsu, 1983). In the control mice, soma of the nerve cells located in the pars compacta of the substantia nigra or SNc (A9 cell group), as well as those in the ventral tegmental area or VTA (A10 cell group), and their processes were intensely stained with anti-TH antibody (Fig. 3A,B,C). In the TIQ-treated mice, however, TH-like immunoreactive (THLI) neurons in the SNc were markedly decreased in number (Fig. 3D,E). The VTA also showed a moderate reduction in the number of THLI neurons (Fig. 3D,F). Although normal-looking THLI neurons could be observed in all of these areas in the TIQ-treated mice, their numbers were markedly reduced. On the other hand, when Cresyl Violet staining was performed, the numbers of neurons in the SNc and VTA of the TIQ-treated mice were almost the same as those of the control mice. In the TIQ-treated mice, some of the nerve cells located in the SNc looked somewhat smaller than those of the control mice. The THLI processes in all areas also remained normal.

In 3 controls and 3 TIQ-treated mice, the number of THLI neurons in the SNc was counted in 20-μm-thick sections processed for TH immunoreactivity. The sections were selected every 200 μm (5 mice) and every 80 μm (1 mouse). THLI neurons were counted for the bilateral SNc in each case. We used the following formula to estimate the total number of THLI neurons: Total counts = sample counts/p, p being the period at which sections were sampled. The mean ± SD of the number of THLI neurons in the SNc was 10190 ± 1812 in control mice (n = 3), while it was 4671 ± 1069 in the TIQ-treated mice, the reduction rate being 56%. TH consists of both active and inactive forms and both forms are stained by this immunohistochemical method. Therefore, dopamine neurons not stained by anti-TH antibody in our experiment were severely damaged in terms of their ability to produce TH protein. However, the neurons themselves were preserved. Thus, we conclude that TIQ did not lead to neuronal death under our experimental conditions. It is known that follow-ing even MPTP administration to young adult mice, recovery from damage occurs both biochemically (Ricaurte et al., 1986; Saitoh et al., 1987) and morphologically (Mori et al., 1988). Since the mice we used were 8-week-old young adults, we must now evaluate the effect of TIQ on aged mice.

N-methyl TIQ

N-methyl TIQ (NMeTIQ, 10 mg/kg, twice a day, S.C.) was administered to 2 aged monkeys (a 14-year-old rhesus and a 15-year-old crab-eating monkey) for 100 days. Their behavioral changes and the neuropathological findings of the substantia nigra were compared to those of 2 control aged monkeys (a 14-

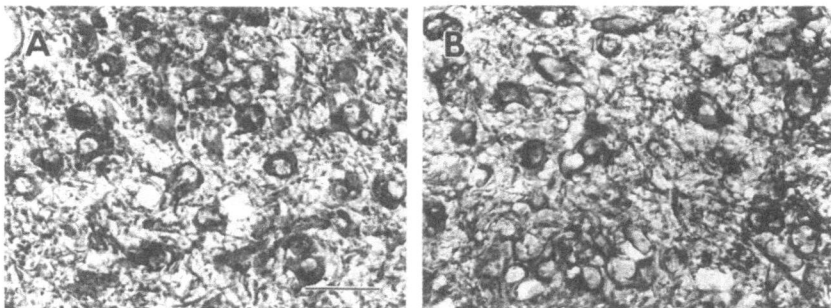

Fig. 4. Tyrosine hydroxylase (TH)-like immunohistochemistry of the pars compacta of the substantia nigra (SNc) of a control monkey (A) and a NMeTIQ-treated monkey (B). TH-like immunoreactive neurons in the SNc of the NMeTIQ-treated monkey are not decreased in number compared to those of the control. Bar = 50μm

year-old rhesus and a 14-year-old crab-eating monkey) who were administered saline for 100 days. Behavioral scores of NMeTIQ-treated monkeys showed slight motor disturbances after 100 days of administration of NMeTIQ, but neuropathological examinations of the substantia nigra revealed no differences between the NMeTIQ-treated and the control monkeys (see Fig. 4).

Conclusion

TIQ has been shown to exist in the brain endogenously and distribute widely in foods. Long-term administration of TIQ produced motor disturbances similar to parkinsonism which were reversed by levodopa, and decreases of concentrations of DA and total BP, and TH activities in the substantia nigra of the monkeys. The similar administration to the mice produced a marked decrease of TH positive neurons of the substantia nigra. The cell death, however, could not be seen in these neurons by Cresyl Violet staining.

Acknowledgment

We are grateful to the support by a Grant-in-Aid for Scientific Research on Priority Areas, Ministry of Education, Science and Culture, Japan.

References

Kohno M, Ohta S, Hirobe M (1986) Tetrahydroisoquinoline and 1-methyl-tetrahydroisoquinoline as novel endogenous amines in rat brain. Biochem Biophys Res Commun 140:448–454

Langston JW, Ballard P, Tetrud JW, Irwin I (1983) Chronic parkinsonism in humans due to a product of meperidine-analog synthesis. Science 219:979–980

Makino Y, Ohta S, Tachikawa O, Hirobe M (1988) Presence of tetrahydroisoquinoline and 1-methyl-tetrahydroisoquinoline in foods: compounds related to Parkinson's disease. Life Sci 43:373–378

Mori S, Fujitake J, Kuno S, Sano Y (1988) Immunohistochemical evaluation of the neurotoxic effects of 1-methyl-4-phenyl-1,2,3,6-tetrahydropyridine (MPTP) on dopaminergic nigrostriatal neurons of young adult mice using dopamine and tyrosine hydroxylase antibodies. Neurosci Lett 90:57–62

Nagatsu I (1983) Immunohistochemistry of biogenic amines and immunoenzyme-histocytochemistry on catecholamine-synthesizing enzymes: application for axoplasmic transport and neuronal localization. In: Methods of biogenic amine research. Elsevier, Amsterdam, pp 873–909

Nagatsu T, Hirata Y (1986) Inhibition of the tyrosine hydroxylase system by MPTP, 1-methyl-4-phenylpyridinium ion (MPP^+) and the structurally related compounds in vitro and in vivo. Eur Neurol 26 [Suppl 1]:11–15

Nagatsu T, Yoshida M (1988) An endogenous substance of the brain, tetrahydroiso-quinoline, produces parkinsonism in primates with decreased dopamine, tyrosine hydroxylase and biopterin in the nigrostriatal regions. Neurosci Lett 87:178–182

Niwa T, Takeda N, Kaneda N, Hashizume Y, Nagatsu T (1987) Presence of tetrahydroisoquinoline and 2-methyl-tetrahydroisoquinoline in parkinsonian and normal human brains. Biochem Biophys Res Commun 144:1084–1089

Niwa T, Takeda N, Tatematsu A, Matsuura S, Yoshida M, Nagatsu T (1988) Migration of tetrahydroisoquinoline, a possible parkinsonia neurotoxin, into monkey brain from blood as proved by gas chromatography-mass spectrometry. J Chromatogr 452:85–91

Niwa T, Yoshizumi H, Tatematsu A, Matsuura S, Nagatsu T (1989) Presence of tetrahydroisoquinoline, a parkinsonism-related compound, in foods. J Chromatogr 493:345–352

Ogawa M, Araki M, Nagatsu I, Nagatsu T, Yoshida M (1989) The effect of 1,2,3,4-tetrahydroisoquinoline (TIQ) on mesencephalic dopaminergic neurons in C57BL/6J mice: immunohistochemical studies-tyrosine hydroxylase. Biogenic Amines 6:427–436

Ohta S, Kohno K, Makino Y, Tachikawa O, Hirobe M (1987) Tetrahydroisoquinoline and 1-methyl-tetrahydroisoquinoline are present in the human brain: relation to Parkinson's disease. Biomed Res 8:453–456

Ricaurte GA, Langston JW, Delanney LE, Irwin I, Peroutka SJ, Forno LS (1986) Fate of nigrostriatal neurons in young mature mice given 1-methyl-4-phenyl-1,2,3,6-tetrahydropyridine: a neurochemical and morphological reassessment. Brain Res 376:117–124

Saitoh T, Niijima K, Mizuno Y (1987) Long-term effect of 1-methyl-4-phenyl-1,2,3,6-tetrahydropyridine (MPTP) on striatal dopamine content in young and mature mice. J Neurol Sci 77:229–235

Yoshida M, Niwa T, Nagatsu T (1990) Parkinsonism in monkeys produced by chronic administration of an endogenous substance of the brain, tetrahydroisoquinoline: the behavioral and biochemical changes. Neurosci Lett 119:109–113

Correspondence: Prof. M. Yoshida, Department of Neurology, Jichi Medical School, Tochigiken, Japan 329-04.

Relationship between TIQs and Parkinson's disease

S. Ohta, M. Kohno, Y. Makino, O. Tachikawa, Y. Tasaki, K. Kikuchi,
and M. Hirobe

Faculty of Pharmaceutical Sciences, University of Tokyo, Japan

Summary

1,2,3,4-Tetrahydroisoquinoline (TIQ) has been assumed to be one of the endo-genous substances inducing Parkinson's disease because of its structural similarity to 1-methyl-4-phenyl-1,2,3,6-tetrahydropyridine (MPTP).

Actually TIQ and its related compound, 1-methyl-1,2,3,4-tetrahydroisoquinoline (1MeTIQ) were coexistent as endogenous amines in rat brain and human brain. 1MeTIQ content was markedly reduced in the parkinsonian brain particularly in the frontal lobe. The 1MeTIQ content also decreased with aging.

TIQs were present in foods, such as wine, cocoa, cheese, milk, egg and banana. In addition, TIQs could penetrate the blood-brain barrier, and hence TIQ from foods may accumulate in the brain over long periods.

Metabolism of TIQ was investigated in connection with metabolism of debrisoquine which is a common marker for the metabolism in human. The main metabolite of TIQ was 4-hydroxy-TIQ. Urinary excretion of 4OH-TIQ was significantly reduced in female DA rat, an animal model of a poor debrisoquine metabolizer. The female DA rat also showed significantly higher brain accumulation of TIQ. These results suggest that the metabolic detoxication process is depressed and TIQ accumulation in the brain is enhanced in a poor debrisoquine metabolizer, which may be one possible explanation for poor debrisoquine metabolizers being susceptible to Parkinson's disease.

Introduction

Since 1-methyl-4-phenyl-1,2,3,6-tetrahydropyridine (MPTP) was found to cause neurotoxicity which results in a clinical syndrome undistinguishable from idiopathic Parkinson's disease in man (Langston et al., 1983), special attention has been paid to the environmental toxic agents in leading selective degenera-tion of the nervous system.

Intake of β-N-oxarylamino-L-alanine (BOAA) in "chickling pea" and β-N-methylamino-L-alanine (BMAA) in cycad seed causes neurodegenerative disorder, lathyrism and amyotropic lateral sclerosis-parkinsonism-dementia complex, respectively (Spencer et al., 1986, 1987).

MPTP is a purely synthetic compound in laboratories, so it cannot be the real environmental agent of idiopathic Parkinson's disease. The real parkinsonism-inducing substance is now sought by many groups focussing on analogous compounds of MPTP.

1,2,3,4-Tetrahydroisoquinoline (TIQ) has been assumed to be one of the endogenous substances inducing Parkinson's disease because of its structural similarity to MPTP. Indeed, parkinsonism can be induced in primates by the administration of TIQ (Nagatsu and Yoshida, 1988).

Here we report the detection of TIQs as endogenous amines and as exogenous compounds in foods. We also mention that change of TIQs contents appear with Parkinson's disease, with aging and with deficiency of drug-metabolizing enzymes.

Endogenous TIQs

Many tetrahydroisoquinoline alkaloids have been found in plants as well as in animals. For example, salsolinol (6,7-dihydroxy-1-methyl-1,2,3,4-tetrahydroisoquinoline) exists in rat brain (Nesterick and Rahwan, 1979) and in human urine as L-dopa metabolite (Sandler et al., 1973). It is suggested that these alkaloids play an important role as neuromodulators and may contribute to pathological conditions such as alcoholism and phenylketonuria (Lasala and Coscia, 1979). As for the structure, almost all tetrahydroisoquinoline alkaloids have electron-donating substituents on the phenyl ring. Because there is no electron-donating group on the ring, TIQ cannot be called a common tetrahydroisoquinoline alkaloid, whereas TIQ and its related compound, 1-methyl-1,2,3,4-tetrahydroisoquinoline (1MeTIQ), were coexistent as endogenous amines in nontreated rat brain by coupled gas chromatography-selected ion monitoring (Kohno et al., 1986). TIQ and 1MeTIQ may be formed by the condensation of phenylethylamine (PEA) with formaldehyde and acetaldehyde, respectively (Fig. 1).

In human brain presenting parkinsonians', TIQ and 1MeTIQ were also present in two regions (frontal lobe and caudate nucleus) (Ohta et al., 1987). The existence of TIQ was also confirmed by Niwa et al. (1987). The TIQ

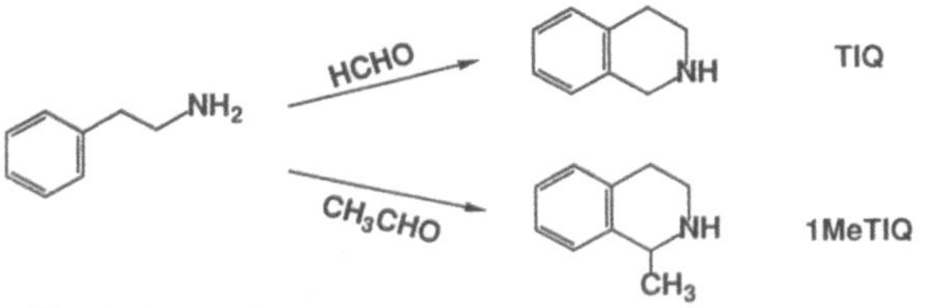

Fig. 1. Proposed mechanism of TIQ and 1MeTIQ formation

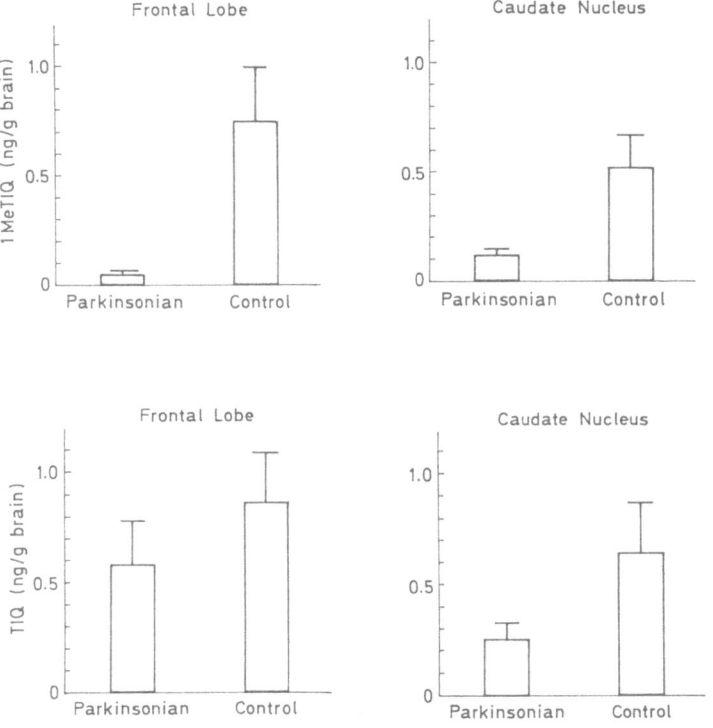

Fig. 2. 1MeTIQ and TIQ contents in brain regions of control and parkinsonian patients (Ohta et al., 1987)

content did not significantly differ between control and parkinsonian brains in both regions (Ohta et al., 1987). In contrast, the 1MeTIQ content was markedly reduced in the parkinsonian brain particularly in the frontal lobe (Fig. 2) (Ohta et al., 1987). Figure 3 shows that the aging process was also an important factor for the content of 1MeTIQ. The content decreased with aging for both the control and the parkinsonian cases (Ohta et al., 1987).

Exogenous TIQs

TIQs can be accumulated if these amines are present in environmental factors, particularly foods. TIQ and 1MeTIQ exist in wine, cocoa and cheese, which are all known as foods with a high PEA content (Table 1) (Makino et al., 1988). TIQ is also present in milk, egg and banana (Niwa et al., 1989). TIQs are thought to be formed by condensation of PEA with aldehydes during ripening and storage of foods.

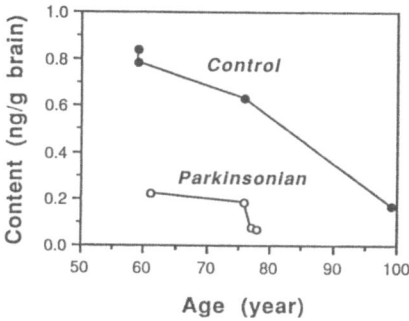

Fig. 3. Relationship between aging and 1MeTIQ content in human brain (Ohta et al., 1987)

Table 1. Concentrations of PEA, 1MeTIQ and TIQ in foods

Sample	1MeTIQ	TIQ	PEA
White wine	354 ± 236	1.7 ± 0.8	$3.5 \times 10^3 \pm 630$
Cheese (Emmenthal)	0.5 ± 0.1	15.0 ± 2.2	240 ± 50
Cocoa A	6.5 ± 0.2	0.8 ± 0.3	670 ± 55
Cocoa B	12.0 ± 2.4	1.1 ± 0.8	$3.1 \times 10^3 \pm 520$

Results are expressed as ng/g of sample. Makino et al. (1988)

TIQs were able to penetrate the blood-brain barrier, and to accumulate in the brain at about 4.5-fold higher concentration than the blood concentration at 4 hr after dosing (Kikuchi et al., 1991). Although the amounts of TIQs in foods are not very high, TIQ from foods may accumulate in the brain over long periods.

Metabolism of TIQ

As mentioned above, the content of TIQ is very small (0.25 – 0.9 ng/g brain) in human brain (Ohta et al., 1987). On the other hand, it is reported that a rather large amount of TIQ administration (50 mg/Kg per day, for 16 days) in primate can produce parkinsonism (Nagatsu and Yoshida, 1988). Since there is a large difference between the content in the brain and the dosage, it may be said that the metabolism (detoxication) of TIQ can be very important.

There is an interesting hypothesis that significantly more parkinsonian than control subjects exhibiting partially or totally defective metabolism of debrisoquine tend to manifest earlier onset of the disease (Barbeau et al., 1985).

Debrisoquine, an antihypertensive agent, is known to be metabolized to 4- hydroxydebrisoquine and is often used as a marker compound of metabolism

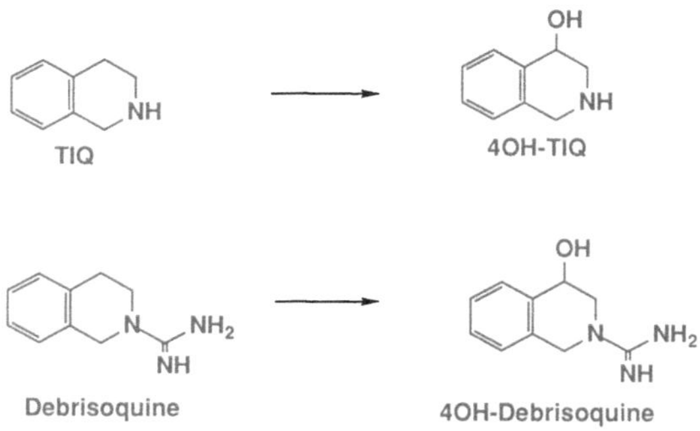

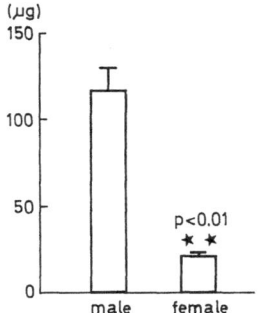

Fig. 4. Metabolism of debrisoquine and TIQ

in humans. It is interesting to note that debrisoquine is structurally very similar to TIQ. Therefore, the metabolism of TIQ seems to be very important from the standpoint of detoxication and the relationship with debrisoquine metabolism.

The main metabolite of TIQ was 4-hydroxy-TIQ, which was the same type of metabolic pathway of debrisoquine, in rat liver microsomes and rat urine (Fig. 4) (Ohta et al., 1990). Hence the correlation between TIQ metabolism and debrisoquine polymorphism was investigated. A possible animal model for the debrisoquine polymorphism has been described in which the female DA rat strain is a poor metabolizer model while the male DA rat strain is an extensive metabolizer model (Gonzalez et al., 1987). Urinary excretion of TIQ metabolite (4OH-TIQ) was significantly lower in female DA rat than in the male rat (Fig. 5). The poor debrisoquine metabolizer rat can effectively metabolize neither debrisoquine nor TIQ.

The brain accumulation and the plasma concentration of TIQ using the same TIQ-treated DA rats were examined, and significantly higher levels were found in female DA rats than in controls (Fig. 6).

Fig. 5. Urinary excretion of 4OH-TIQ 24 hr after oral administration of TIQ to DA rat (Ohta et al., 1990)

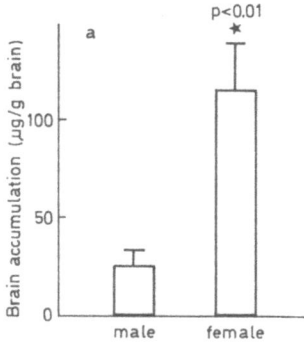 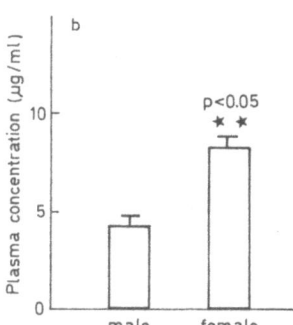

Fig. 6. a Brain accumulation and b plasma concentration of TIQ 24 hr after oral administration of TIQ to DA rat (Ohta et al., 1990)

From the above results, poor debrisoquine metabolizers seem to suppress the metabolism of TIQ, and a significant amount of TIQ may accumulate in the brain over long periods. This process may be one of the mechanisms for the onset of Parkinson's disease.

Conclusions

The mechanism of the onset of Parkinson's disease has not yet been determined; therefore, it is necessary to employ many approaches to this problem.

It has become apparent that TIQ, a possible parkinsonism-inducing substance, exists in rat as well as human brain, as well as in some foods. The content of 1MeTIQ in the human brain appears to be markedly decreased in parkinsonian. Therefore, 1MeTIQ is a candidate of a marker for parkinsonism and aging. It is frequently observed that the aging process is involved in inducing parkinsonism. In this sense, 1MeTIQ might plays a role in protecting the brain from the parkinsonism or aging process. It can be said that the pharmacological effects of 1MeTIQ are very important; thus, these effects are now under investigation.

References

Barbeau A, Cloutier T, Roy M, Plasse L, Paris S, Poirier J (1985) Ecogenetics of Parkinson's disease: 4-hydroxylation of debrisoquine. Lancet ii:1213–1216

Gonzalez FJ, Matsunaga T, Nagata K, Meyer UA, Nebert DW, Pastewka J, Kozak CA, Gillette J, Gelboin HV, Hardwick JP (1987) Debrisoquine 4-hydroxylase: characterization of a new P450 gene subfamily, regulation, chromosomal mapping, and molecular analysis of the DA rat polymorphism. DNA 6:149–161

Kikuchi K, Nagatsu Y, Makino Y, Mashino T, Ohta S, Hirobe M (1991) Metabolism and penetration through blood-brain barrier of parkinsonism-related compounds:

1,2,3,4-tetrahydroisoquinoline (TIQ) and 1-methyl-1,2,3,4-tetrahydroisoquinoline (1 MeTIQ). Drug Metab Dispos 19:257–262

Kohno M, Ohta S, Hirobe M (1986) Tetrahydroisoquinoline and 1-methyltetrahydroisoquinoline as novel endogenous amines in rat brain. Biochem Biophys Res Commun 140:448–454

Langston JW, Ballard P, Tetrud JW, Irwin I (1983) Chronic parkinsonism in humans due to a product of meperidine-analog synthesis. Science 219:979–980

Lasala JM, Coscia CJ (1979) Accumulation of a tetrahydroisoquinoline in phenylketonuria. Science 203:283–284

Makino Y, Ohta S, Tachikawa O, Hirobe M (1988) Presence of tetrahydroisoquinoline and 1-methyltetrahydroisoquinoline in foods: compounds related to Parkinson's disease. Life Sci 43:373–378

Nagatsu T, Yoshida M (1988) An endogenous substance of the brain, tetrahydroisoquinoline, produces parkinsonism in primates with decreased dopamine, tyrosine hydroxylase and biopterin in the nigrostriatal regions. Neurosci Lett 87:178–182

Nesterick CA, Rahwan RG (1979) Detection of endogenous salsolinol in neonatal rat tissue by a radioenzymatic-thin-layer chromatographic assay. J Chromatogr Biomed Appl 164:205–216

Niwa T, Takeda N, Kaneda N, Hashizume Y, Nagatsu T (1987) Presence of tetrahydroisoquinoline and 2-methyl-tetrahydroquinoline in parkinsonian and normal human brains. Biochem Biophys Res Commun 144:1084–1089

Niwa T, Yoshizumi H, Tatematsu A, Matsuura S, Nagatsu T (1989) Presence of tetrahydroisoquinoline, a parkinsonism-related compound, in foods. J Chromatogr Biomed Appl 493:347–352

Ohta S, Kohno M, Makino Y, Tachikawa O, Hirobe M (1987) Tetrahydroisoquinoline and 1-methyltetrahydroisoquinoline are present in the human brain: relation to Parkinson's disease. Biomed Res 8:453–456

Ohta S, Tachikawa O, Makino Y, Tasaki Y, Hirobe M (1990) Metabolism and brain accumulation of tetrahydroisoquinoline (TIQ), a possible parkinsonism-inducing substance, in an animal model of a poor debrisoquine metabolizer. Life Sci 46:599–605

Sandler M, Carter SB, Hunter KR, Stern GM (1973) Tetrahydroisoquinoline alkaloids: in vivo metabolites of L-dopa in man. Nature 241:439–443

Spencer PS, Roy DN, Ludolph A, Hugon J, Dwivedi MP, Schaumburg HH (1986) Lathyrism: evidence for role of the neuroexcitatory amino acid BOAA. Lancet ii:1066–1067

Spencer PS, Nunn PB, Hugon J, Ludolph AC, Ross SM, Roy DN, Robertson RC (1987) Guam amyotrophic lateral sclerosis-parkinsonism-dementia linked to a plant excitant neurotoxin. Science 237:517–522

Correspondence: Dr. S. Ohta, Faculty of Pharmaceutical Sciences, University of Tokyo, Hongo, Bunkyo-ku, Tokyo 113, Japan

Sparteine oxidation in patients with early- and late-onset of Parkinson's disease

K. Chiba[1], H. Imai[2], H. Yoshino[2], J. Kato[1], T. Ishizaki[1], and H. Narabayashi[2]

[1] Division of Clinical Pharmacology, Clinical Research Institute, National Medical Center, and
[2] Department of Neurology, Juntendo University School of Medicine, Tokyo, Japan

Summary

We studied the oxidation of sparteine in 36 patients with early onset (onset of age <50 years) and 36 patients with late onset (onset of age ≥50 years) of Parkinson's disease. The results were compared with those obtained from 84 healthy volunteers. Although no poor metabolizer was found in the patients with Parkinson's disease, there was an inverse and significant (P < 0.01) relationship between the onset of age and log metabolic ratio (MR) of sparteine (Spearman's $r_s = -0.431$). The mean (±S.D.) MR value (1.52 ± 1.40) of the patients with early onset was significantly (P < 0.01) higher than, whereas that (0.81 ± 0.99) of the patients with late onset was not significantly different from, that (0.62 ± 0.62) of the control subjects with extensive metabolizer phenotype. The mean greater MR value found in the patients with early onset of Parkinson's disease was derived mainly from the more frequency of patients with relatively greater MR value for phenotyping extensive metabolizers (MR > 1.5). These findings indicate that the oxidative capacity of sparteine is not defective but reduced in some patients with an earlier onset of Parkinson's disease, possibly implying that the decreased activity of cytochrome P-450 responsible for the oxidation of sparteine would be a factor to accelerate the onset of the Parkinson's disease.

Introduction

Since the discovery of 1-methyl-4-phenyl-1,2,3,6-tetrahydropyridine (MPTP) which causes parkinsonism-like syndrome in humans and monkeys, neurotoxin(s) with a chemical structure similar to MPTP in the environments has been suggested as a causative substance of idiopathic Parkinson's disease (Langston and Irwin, 1986). Because many toxic chemicals are detoxified by the hepatic microsome cytochrome P-450, people with a low activity or defect in the enzyme may be predisposed to idiopathic Parkinson's disease. Barbeau and his coworkers (1985) first tested this hypothesis using debrisoquine as the probe drug which is metabolized by one of the P-450 isozymes, called as P-450 IID6 or P-450 dbl (Brosen, 1990; Eichelbaum and Gross, 1990). The metabolism of debrisoquine is characterized by genetically determined, two phenotypes, extensive (EMs) and poor metabolizers (PMs). They found that significantly more patients with Parkinson's disease than the control subjects have partially or totally defective 4-hydroxylation of debrisoquine (Barbeau et al., 1985). They also found that PMs of debrisoquine are patients with an earlier onset of the disease (Barbeau et al., 1985).

Later, the same group retracted some parts of these findings and attributed the lower oxidative rate of debrisoquine to an inhibitory effect of concurrently administered antihistamine drugs (Poirier et al., 1987). Other investigators reported that there is no difference in the frequency of PMs of debrisoquine between the patients with Parkinson's disease and the control subjects (Steventon et al., 1989; Gudjonsson et al., 1990). However, Poirier et al. (1987) stated in their retracted report that the observation of an earlier age of the onset of Parkinson's disease in PMs of debrisoquine still holds true. Recently, Benitez et al. (1990) also reported that there is a negative correlation between the metabolic capacity of debrisoquine and onset of Parkinson's disease.

In view of the reports (Poirier et al., 1987; Benitez et al., 1990) suggesting that the oxidative metabolism of debrisoquine is somewhat altered in the patients with the earlier onset of Parkinson's disease, we compared the oxidation capacity of sparteine, which is cosegregated with the metabolism of debrisoquine (Eichelbaum et al., 1982; Eichelbaum and Gross, 1990), between patients with the early onset (<50 years old) and those with the late onset of Parkinson's disease (≥ 50 years old).

Materials and methods

Patients and control subjects

Seventy-two patients with idiopathic Parkinson's disease who attended the Neurology Clinic at the Juntendo University Hospital (Tokyo, Japan) were included in the study. Informed consent was obtained from each of the patients. Patients taking drug(s) known to affect sparteine metabolism (e.g., neuroleptics, tricyclic antidepressants, beta blockers, quinidine, cimetidine; Eichelbaum and Gross, 1990) were excluded. Patients

Table 1. Characteristics of the patient subgroups

Group	Number of patients	Sex (M/F)	Onset of age (years)	Present age (years)
Subgroup I	36	17/19	37.3 ± 7.7^a (16−49)	53.6 ± 9.1^a (35−74)*
Subgroup II	36	21/15	61.3 ± 7.0 (50−76)	67.5 ± 7.4 (53−82)

[a] Significantly ($P < 0.01$) different from the subgroup II. * Ranges included

with hepatic or ranal disease were also excluded. No patients received any antihistamine drug or were chronic ethanol intakers.

The age of the onset of the movement disorder was documented for each patient by interviewing the patient and his or her relative who lived with the patient. Patients with an age of the onset of <50 years at developing the motor symptoms of Parkinson's disease were classified arbitrarily into the early onset group (subgroup I), whereas those with an age of the onset of ≥50 years were difined as the late onset group (subgroup II). The demographic characteristics of the patients are listed in the Table 1. Thirty-six patients formed each of the subgroups. The mean ages of the onset of the disease were 37.3 years and 61.3 years for the subgroups I and II, respectively. The control group consisted of 84 healthy Japanese volunteers (10 females, 74 males, mean age, 27.4 ± years) reported previously (Ishizaki et al., 1987).

Phenotype determination

After emptying his or her bladder, each took by mouth 100 mg of sparteine sulfate (Depasan® tablet, Giulini Pharma, Hannover, West Germany) following overnight fast. During the following 6 hours, urine was collected, the volume measured, and an aliquot was stored at −20°C until analyzed.

Sparteine and its metabolites, 2- and 5-dehydrosparteine, were determined by the method of Eichelbaum et al. (1982). PMs of sparteine were defined as subjects with a urinary metabolic ratio (MR) greater than 20 (Eichelbaum et al., 1982), which was calculated from: amount of sparteine / amount of 2- plus 5-dehydrosparteine.

Statistics

The data are given as mean ± S.D. Statistical analysis was performed using ANOVA, Chi-squared test, and Spearman's rank correlation where appropriate. A P value of less than 0.05 was considered statistically significant.

Results

No patient was classified as PM in both of the patient subgroups, whereas two of the 84 control subjects had a MR >20 and thus were judged as PMs (Table 2). Although the frequency of PMs was not higher in the patients with Parkinson's disease than that in the controls, the mean MR value in the subgroup I was 2.5 times greater than that of the control subjects when the mean value was compared between the subgroup I patients and control subjects with EM phenotype (Table 2). The mean value was also approximately two times greater in the subgroup I than in the subgroup II (Table 2). The differences in the MR values between the subgroup I and control EM subjects, and between the subgroups I and II were statistically significant (P < 0.01). Although the mean MR value of the subgroup II patients was slightly higher

Table 2. Frequency of poor metabolizers (PMs) of sparteine oxidation in patients and control subjects, and metabolic ratio (MR) and frequency of extensive metabolizers (EMs) with MR of greater than 1.5 in patients and control subjects with EM phenotype

Group	Number of subjects	PMs	MR of EMs	EMs with MR > 1.5 /total EMs
Patients				
Subgroup I	36	0	$1.52 \pm 1.40^{a,b}$ (0.28−6.12)*	$11/36^{a,b}$
Subgroup II	36	0	0.81 ± 0.99 (0.02−5.99)	4/36
Controls	84	2	0.62 ± 0.62 (0.11−3.07)	7/82

[a] Significantly ($P < 0.01$) different from the control group. [b] Significantly ($P < 0.01$) different from the subgroup II. * Ranges included

than that of the control subjects with EM phenotype, the difference was not statistically significant (Table 2). In all the patients studied, there was an inverse and significant relationship between the onset of age and log MR ($r_s = -0.431$, $P < 0.01$, $n = 72$).

Since the differences in the mean MR values among the patient subgroups and control group appeared to be due to the greater frequency of the patients with a relatively higher MR value in the subgroup I, we examined the frequency of MR values being greater than 1.5 in the patient and control groups. The cutoff point of the value of 1.5 was chosen since it is the lower limit of 95% confidence for the MR of heterozygotes found in the family study of one PM in the control group (Chiba et al., 1988a). The frequency of EMs with a MR of > 1.5 was significantly ($P < 0.05$) greater in the subgroup I than in the subgroup II and in the control EM subjects. There was no significant difference in such a trend between the subgroup II EM patients and control EM subjects.

Discussion

We found no PM of sparteine oxidation capacity in our patients although they included 21 patients with juvenile type of Parkinson's disease (onset of age <40 years). This is inconsistent with the report of Barbeau et al. (1985), which showed that all the patients with a phenotype of PM were those with the onset of age less than 40. However, we found that the age at the onset of Parkinson's disease is inversely and significantly correlated with log MR of sparteine, being compatible with the findings of Benitez et al. (1990) using debrisoquine as the test probe. In addition, the mean MR value of the patients with early onset was significantly greater than that of the EM patients with late onset as well as that of the control EM subjects, while the mean MR value of the EM patients with late onset was not statistically different from that of the control EM subjects (Table 2). These findings suggest that the oxidative capacity of sparteine is not defective but reduced in the patients with an earlier onset of Parkinson's disease.

The greater MR of sparteine found in the subgroup I patients appears to be due mainly to the higher frequency of patients with a relatively greater MR value: the frequency of the patients with a MR of larger than 1.5 was significantly higher in the EM patients with early onset than in the EM patients with late onset and in the control EM subjects (Table 2). Since the cutoff value (MR = 1.5) was chosen as the lower limit of 95% confidence for the MR of heterozygotes found in the family study of one PM in the control subjects (Chiba et al., 1988a), our findings suggest that more patients with heterozygous status of sparteine oxidation is likely to be included in the patients with an earlier onset of Parkinson's disease. However, the MR of sparteine is influenced not only by the genetic background but also by the other factors including concurrently administered drugs (Eichelbaum and Gross, 1990). Since we could not withdraw the medication for the treatment of parkinsonism from the patients for the ethical reasons, it might have interfered with the metabolism of sparteine and increased the MR values in our patients. Nonetheless, none of our patients was treated with drugs known to interfere with sparteine metabolism. In fact, there was no significant difference in the mean value of MR between the EM patients with late onset and control EM subjects (Table 2). Furthermore, the pattern of the medications prescribed to the patients (levodopa, carbidopa, benserazide and/or bromocriptine) was not different between the patients with early (subgroup I) and late onsets (subgroup II) of Parkinson's disease. Thus, the increased MR of sparteine found in the patients with early onset does not appear to be due to the medications for the treatment of Parkinson's disease.

As a matter of course, no definite conclusion can be drawn from the present results, regarding the association of heterozygotes of sparteine oxidation capacity with relatively higher MR values found in the patients with early onset of Parkinson's disease, since the heterozygous status cannot be determined by the MR alone. Family studies for the individual patient are obviously required for clarifying this controversial issue. However, the genetic inheritance has been suggested to play an important role in the development of the early onset of Parkinson's disease, since the remarkably greater incidence of the disease has been found in the relatives of the patients with the juvenile type of Parkinson's disease (Yokochi and Narabayashi, 1981). Thus, a possibility may exist that the "slow" allele of gene encoding for P-450 dbl is associated with the development of Parkinson's disease with the early onset of age. Further studies are also needed to clarify this possibility.

Since the discovery of MPTP, the exposure to an environmental toxin with a chemical structure similar to MPTP has been suggested to be responsible for the possible cause of Parkinson's disease (Langston and Irwin, 1986). Many toxic chemicals are detoxified by the hepatic microsomal cytochrome P-450 system. Sparteine is oxidized by one of the isozymes of cytochrome P-450 (i.e., P-450 dbl), which has been reported to be involved in the metabolism of MPTP (Fonne-Pfister and Meyer, 1988). Although the metabolism by the cytochrome P-450 is not a major detoxication pathway of MPTP at least in mice (Chiba et al., 1988b, 1990), it is likely that substances structurally related to MPTP that are metabolized by P-450 dbl would be involved in one of the causes of Parkinson's disease.

Recently, tetrahydroisoquinoline has been reported to decrease dopamine, tyrosine hydroxylase and biopterin in the nigrostriatal regions, and to produce parkinsonism-like syndrome in marmosets (Nagatsu and Yoshida, 1988). The compound exists in our environments as a contaminant of foods such as cheese, wine and cocoa (Makino et al., 1988). This compound is considered to be metabolized by P-450 dbl, since the urinary excretion of 4-hydroxy metabolite of tetrahydroquinoline was reduced in female AD rats, an animal model of debrisoquine hydroxylation deficiency (Ohta et al., 1990). When tetrahydroisoquinoline was administered orally, its concentration in the brain was significantly greater in the female DA rats with debrisoquine hydroxylation deficiency than in the male DA rats with normal activities for the hydroxylation of debrisoquine (Ohta et al., 1990). These findings appear to suggest that a subject with the reduced activity of cytochrome dbl may be more susceptible to the neurotoxic effect of tetrahydroisoquinoline existing in the environments.

However, the present study showed that there were no patients with PM phenotype of sparteine oxidation related to a negligible activity of P-450 dbl in both patient subgroups with the early and late onsets of Parkinson's disease. Based upon this finding, we are tempted to speculate that tetrahydroisoquinoline in the environments might not play a major role in the development of idiopathic Parkinson's disease. Nonetheless, the possibility exists that tetrahydropyridine or its related compound(s) metabolized by P-450 dbl is likely to accelerate the onset of Parkinson's disease for a person who is prone to acquire Parkinson's disease, since our results showed that the capacity of sparteine oxidation is reduced in some of the patients with early onset of Parkinson's disease. As a whole, the reduced activities of P-450 dbl may not be a major factor, but may be one of the possible factors, determining the individual susceptibility to the environmental neurotoxins. This context appears to be in accordance with the generally accepted theory on the multifactorial pathoetiology of Parkinson's disease such that Parkinson's disease is caused by a combination or accumulation of many relatively minor factors (Kondo et al., 1973; Marsden, 1990).

Acknowledgments

The authors wish to thank Mrs. T. Chiba for her secretarial assistance. This work was supported by the grant-in-aid of the Japan Health Science Foundation (3-4-3-B) and the Sankyo Science Foundation.

References

Barbeau A, Cloutier T, Roy M, Plasse L, Paris S, Poirier J (1985) Ecogenetics of Parkinson's disease: 4-hydroxylation of debrisoquine. Lancet ii:1213–1216
Benitez J, Ladero JM, Jimenez-Jimenez FJ, Martinez C, Puerto AM, Valdivielso MJ, Llerena A, Cobaleda J, Munoz JJ (1990) Oxidative polymorphism of debrisoquine in Parkinson's disease. J Neurol Neurosurg Psychiatry 53:289–292
Brosen K (1990) Recent developments in hepatic drug oxidation. Implications for clinical pharmacokinetics. Clin Pharmacokinet 18:220–239

Chiba K, Horii H, Kubota E, Ishizaki T, Kato Y (1990) Effects of N-methylmer-captoimidazole on the disposition of MPTP and its metabolites in mice. Eur J Pharmacol 180:59–67

Chiba K, Kato J, Hashimoto K, Ishizaki T (1988a) Apparent Mendelian recessive inheritance of sparteine metabolism in an extended Japanese family. Eur J Clin Pharmacol 34:661–662

Chiba K, Kubota E, Miyakawa T, Kato Y, Ishizaki T (1988b) Characterization of hepatic microsomal metabolism as an in vivo detoxication pathway of 1-methyl-4-phenyl-1,2,3,6-tetrahydropyridine in mice. J Pharmacol Exp Ther 246:1108–1115

Eichelbaum M, Bertilson L, Säwe J, Zekorn C (1982) Polymorphic oxidation of spar-teine and debrisoquine: related pharmacogenetic entities. Clin Pharmacol Ther 31:184–186

Eichelbaum M, Gross S (1990) The genetic polymorphism of debrisoquine/sparteine metabolism – clinical aspects. Pharmacol Ther 46:377–394

Fonne-Pfister R, Meyer UA (1988) Xenobiotic and endobiotic inhibitors of cytochrome P-450 dbl function, the target of the debrisoquine/sparteine type polymorphism. Biochem Pharmacol 37:3829–3835

Gudjonsson O, Sanz E, Alvan G, Aquilonius S-M, Reviriego J (1989) Poor hydroxylator phenotypes of debrisoquine and S-mephenytoin are not over-represented in a group of patients with Parkinson's disease. Br J Pharmacol 30:301–302

Ishizaki T, Eichelbaum M, Horai Y, Hashimoto K, Chiba K, Dengler HJ (1987) Evidence for polymorphic oxidation of sparteine in Japanese subjects. Br J Clin Pharmacol 23:482–48

Kondo D, Kurland LT, Schull WJ (1973) Parkinson's disease: genetic analysis and evidence of a multifactorial etiology. Mayo Clin Proc 48:465–475

Langston JW, Irwin I (1986) MPTP: current concepts and controversies. Clin Neuro-pharmacol 9:487–507

Makino Y, Ohta S, Tachikawa O, Hirobe M (1988) Presence of tetrahydroisoquinoline and 1-methyl-tetrahydro-isoquinoline in foods: compounds related to Parkinson's disease. Life Sci 43:373–378

Marsden CD (1990) Parkinson's disease. Lancet i:948–952

Nagatsu T, Yoshida M (1988) An endogenous substance of the brain, tetrahydro-isoquinoline, produces parkinsonism in primates with decreased dopamine, tyrosine hydroxylase and biopterin in the nigrostriatal regions. Neurosci Lett 87:178–182

Ohta S, Tachikawa O, Makino Y, Tasaki Y, Hirobe M (1990) Metabolism and brain accumulation of tetrahydroisoquinoline (TIQ). A possible parkinsonism inducing substance, in an animal model of a poor debrisoquine metabolizer. Life Sci 46:599–606

Poirier J, Roy M, Campanella G, Cloutier T, Paris S (1987) Debrisoquine metabolism in Parkinsonian patients treated with antihistamine drugs. Lancet ii:386

Steventon GB, Heafield MTH, Waring RH, Williams AC (1989) Xenobiotic metab-olism in Parkinson's disease. Neurology 39:883–887

Yokochi M, Narabayashi H (1981) Clinical characteristics of juvenile parkinsonism. In: Rose FC, Capildeo R (eds) Research progress in Parkinson's disease. Pitman Medical, Tunbridge Wells, Kent, pp 35–39

Correspondence: Dr. K. Chiba, Division of Clinical Pharmacology, Clinical Research Institute, National Medical Center, 1-21-2 Toyama, Shinjuku-ku, Tokyo 162, Japan.

Association of Xba I allele (Xba I 44kb) of the human cytochrome P-450dbI (CYP2D6) gene in Japanese patients with idiopathic Parkinson's disease

I. Kondo[1] and I. Kanazawa[2]

[1] Departments of Human Ecology and Genetics, Faculty of Medicine, University of the Ryukyus, Okinawa, and [2] Department of Neurology, Institute of Clinical Medicine, University of Tsukuba, Ibaraki, Japan

Summary

Cytochrome P-450dbl gene (CYP2D6) Xba I genotypes were investigated in 43 unrelated healthy Japanese and 51 patients with idiopathic Parkinson's disease (IPD), using genomic DNA hybridization analysis with a cDNA probe encoding cytochrome P-450dbl. Restriction fragment length polymorphisms (RFLPs) were observed in both populations, but the frequencies of alleles detected by restriction enzyme Xba I differed significantly. The relative risk of IPD is 2.15 times greater for idividuals with a Xba I 44 kb allele compared to those without the allele ($x^2 = 4.149$, d.f. = 1, $p < 0.05$). Furthermore, the relative risk of IPD is 6.32 times greater for the Xba I 44 kb homozygotes than the Xba I 29 kb homozygotes ($x^2 = 4.935$, d.f. = 1, $p < 0.05$). These data suggest that the Xba I 44 kb allele of CYP2D6 is one of the genetic factors making humans susceptible to IPD acquisition.

Introduction

Idiopathic Parkinson's disease (IPD) is thought to be a multifactorial disorder and is a result of environmental factors acting on genetically susceptible individuals with normal aging (Barbeau et al., 1986). The findings of Parkinsonism-induced neurotoxin, 1-methyl-4-phenyl-1,2,3,6-tetrahydropyridine (MPTP) has led to speculation that similar toxins in the environment may be responsible for the pathogenesis of IPD, and recent epidemiological studies have focused on searching for clues to possible environmental factors (Langston et al., 1983). MPTP is metabolized in the liver by cytochrome P-450dbl (P-450dbl) and causes a marked depression of the hepatic P-450dbl monooxygenase system (Shahi et al., 1989). Among the alkaloids and drugs, over 25 were also competitive inhibitors of the P-450dbl function (Fonne-Pfister and Meyer, 1988).

Genetic polymorphisms of these drug oxidation have been studied and two distinct phenotypes, extensive metabolizer (EM) and poor metabolizer (PM), have been identified (Mahgoub et al., 1977). PM phenotype behaves as an autosomal recessive trait. Barbeau et al. (1985) reported that significantly more patients with IPD than control subjects were partially or totally defective in debrisoquine metabolism. At present, debrisoquine metabolism is being studied by the administration of a test drug (debrisoquine, sparteine, or dextromethorphan) and the examination of the ratio between drug and metabolite in urine. However, the procedure is difficult to study in a large population, and heterozygous carries for the PM gene have not been identified. Identification of other genetic markers related to P-450dbl would be more convenient in studying the association of the genetic factor with the pathogenesis of IPD.

Recently, Gonzalez et al. (1988) cloned human P-450dbl complementary DNA, designated CYP2D6, from liver and restriction fragment length polymorphisms (RFLPs) with CYP2D6 have been identified in Caucasians (Skoda et al., 1988) and in Chinese (Yue et al., 1989). We investigated CYP2D6 polymorphisms using genomic DNA hybridization analysis in Japanese patients with IPD. Association analysis between genotypes of CYP2D6 and IPD was performed in 2 × 2 tables by chi-square statistics.

Materials and methods

Peripheral blood was collected from 43 healthy members of the university staff and students (age range 19–60) and 51 patients with IPD (age range 26–78) attending the neurology clinics at Okinawa Kyoudou Hospital, Naha City Hospital and Okinawa Prefectural Hospital in Okinawa, located in south-western Japan. DNA was isolated from PHA-stimulated lymphocytes according to the method of Kunkel et al. (1977). Five μg of DNA were digested with Xba I under conditions recommended by the manufacturer (Nippon Gene Co., Ltd., Toyama, Japan). Digests of DNA were subjected to electrophoresis in 0.8% agarose gels and transferred to nylon membrane filters (Dupont, MA, USA) by Southern blotting (Southern, 1975). The filters were hybridized with a ^{32}P-labelled CYP2D6 probe, kindly provided by Dr. F.J. Gonzalez, which was a 1.6 kb fragment containing the entire coding region of the P-450dbl gene (Gonzalez et al., 1988). Autoradiographs were developed at −70°C using Kodak XAR-5 film and superrapid intensifying screens for three days.

Association analysis between the genotypes identified by the CYP2D6 polymorphisms and IPD was performed in a 2 × 2 table by chi-square statistics.

Results

Figure 1 shows an autoradiogram of the CYP2D6 Xba I genotypes in four individuals. Xba I-digested DNA defined four allelic forms of the gene for P450db1: the most common allele is characterized by the presence of a 29 kb

Table 1. Distribution of Xba I CYP2D6 genotypes in healthy subjects and patients with idiopathic Parkinson's disease (IPD)

Xba I CYP2D6 Genotype	Healthy subjects		x^2	Patients with IPD		x^2
	Observed No.	Expected No.		Observed No.	Expected No.	
44 kb	1	2.1	0.58	6	2.5	4.90
44 kb − 29 kb	17	13.7	0.79	24	16.3	3.64
44 kb − 11.5 kb	0	0.9	0.90	1	1.1	0.00
44 kb − 16 kb + 9 kb	0	0.2	0.20	0	0.3	0.30
29 kb	20	22.4	0.26	19	26.5	2.12
29 kb − 11.5 kb	4	2.9	0.42	1	3.5	1.79
29 kb − 16 kb + 9 kb	1	0.7	0.13	0	0.9	0.90
11.5 kb	0	0.1	0.10	0	0.1	0.10
11.5 kb − 16 kb + 9 kb	0	0.0	0.00	0	0.0	0.00
16 kb + 9 kb − 16 kb + 9 kb	0	0.0	0.00	0	0.0	0.00
Total	43	43.0	3.38	51	51.2	13.75

p44 kb = 0.221, p29 kb = 0.721, p11.5 kb = 0.047, p16 kb + 9 kb = 0.012

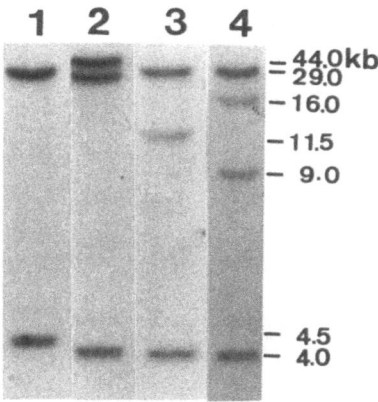

Fig. 1. Hybridization patterns of CYP2D6 probe to Xba I-digested DNAs from four individuals. *Lane 1* Homozygote with a Xba I 29 kb allele. *Lane 2* Heterozygote with a Xba I 29 kb allele and a Xba I 44 kb allele. *Lane 3* Heterozygote with a Xba I 29 kb allele and a Xba I 11.5 kb allele. *Lane 4* Heterozygote with a Xba I 29 kb allele and a Xba I 16 kb and 9 kb allele

fragment (Xba I 29 kb) (Fig. 1, lane 1); a second allele is identified by the presence of a 44 kb fragment (Xba I 44 kb) (Fig. 1, lane 2); a third allele is defined by the presence of an 11.5 kb fragment (Xba I 11.5 kb) (Fig. 1, lane 3); and a fourth allele is determined by the presence of 16 kb and 9 kb fragments (Xba I 16 kb + 9 kb) (Fig. 1, lane 4).

The distribution and allele frequencies of CYP2D6 Xba I genotypes in 43 unrelated healthy Japanese are summarized in Table 1. In healthy individuals, the observed numbers are close to those calculated on the assumption that the Hardy-Weinberg equilibrium persists ($x^2 = 3.38$, d.f. = 6, $0.75 < p < 0.09$). On the other hand, the distribution of CYP2D6 genotypes in 51 patients with IPD differ from those in healthy subjects, and observed numbers of genotypes are significantly different from those calculated on the assumption that the Hardy-Weinberg equilibrium persists ($x^2 = 13.75$, d.f. = 6, $0.025 < p < 0.05$).

Table 2. Distribution of homozygotes of Xba I CYP2D6 genotype in healthy subjects and patients with IPD

Xba I CYP2D6 Genotype	Healthy subjects	Patients with IPD	Total
44 kb homozygote	1	6	7
29 kb homozygote	20	19	39
Total	21	25	46

Odds ratio (0) = 6.32 x^2 = 4.935 > x^2 0.05 = 3.841 (4.510 ~ 8.129)

Table 3. Distribution of Xba I 44 kb genotype of CYP2D6 in healthy subjects and patients with IPD

Xba I CYP2D6 Phenotype	Healthy subjects	Patients with IPD	Total
44 kb presence	18	31	49
44 kb absence	25	20	45
Total	43	51	94

Odds ratio (0) = 2.15 x^2 = 4.149 > x^2 0.05 = 3.841 (0.323 ~ 3.979)

Table 2 shows the distribution of the Xba I 44 kb and 29 kb homozygotes in healthy subjects and in patients with IPD. The relative risk of IPD is 6.32 times more for Xba I 44 kb homozygotes than for Xba I 29 kb homozygotes ($x^2 = 4.935$, d.f. = 1, p < 0.05). In addition, as shown in Table 3, the distribution of presence or absence of the Xba I 44 kb allele at the CYP2D6 locus in healthy individuals and patients with IPD is also different. The relative risk of IPD is 2.15 times greater for individuals with the Xba I 44 kb allele compared to those without the allele ($x^2 = 4.149$, d.f. = 1, p < 0.05).

Discussion

Our findings suggest that the relative risk of IPD differs between individuals with different Xba I genotypes of CYP2D6. The individuals with one or two Xba I 44 kb alleles were 2.15 times more susceptible to IPD acquisition compared to individuals without the Xba I 44 kb allele. Furthermore, the relative risk of IPD in Xba I 44 kb homozygotes was 6.32 times more likely compared with that in Xba I 29 kb homozygotes. These data suggest that the Xba I 44 kb allele on the CYP2D6 locus might have an impact on individual susceptibility to IPD in humans.

By RFLPs analyses, the associations of several mutant alleles of the CYP2D6 gene locus with the poor metabolizer (PM) has been identified in Caucasians (Skoda et al., 1988; Heim and Meyer, 1990), and the Xba I 44 kb and 11.5 kb alleles were in linkage disequilibrium with the PM phenotype in Caucasians. The frequencies of these two alleles were 0.221 and 0.047 in healthy Japanese population. Then, a respective number of individuals with the defective alleles of CYP2D6 (Xba I 44 kb and 11.5 kb) may be about 7.2% calculated on gene frequencies in Japanese. On the other hand, based on studies of debrisoquine metabolism, the frequency of PM was about 3% in the Japanese population (Ishizaki et al., 1987), and Nakamura et al. (1985) could not identify the PM in 100 unrelated healthy Japanese. Therefore, the Xba I 44 kb genotype may represent either an active or a defective allele in Japanese.

Recently, Yue et al. (1989) reported that five Chinese homozygotes for the Xba I 44 kb had the EM debrisoquine hydroxylation phenotype, and Xba I genotype was discordant in the Chinese studied. However, their data seemed

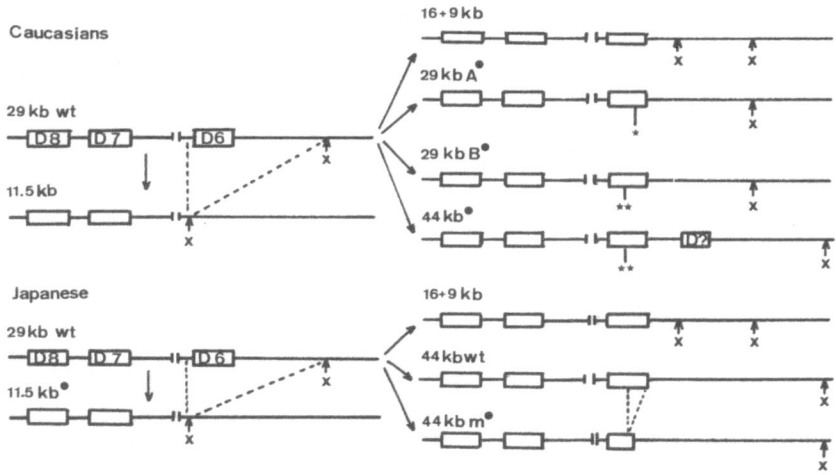

Fig. 2. Structure of CYP2D6-8 gene cluster in Caucasians and Japanese
X Xba I restriction site; * nucleotide deletion; ** splice side defect; 44 kb m* defective allele of P-450dbl function

to suggest that individuals with the Xba I 44 kb allele were likely to be PM carriers because the frequencies of the Xba I 44 kb allele differed within two small populations. In 8 individuals with the metabolic ratio (MR) of debrisoquine hydroxylation of 2–12.6, seven had the Xba I 44 kb allele. However, 5 individuals in 13 with the MR of <2 had not the Xba I 44 kb allele. Thus, the Xba I 44 kb allele with defective cytochrome P-450dbI function may present in the Chinese and Japanese populations. Figure 2 shows the structure of the CYP2D6-8 gene cluster in Caucasians (Heim and Meyer, 1990) and Japanese. The evolution of mutant alleles of CYP2D6 may be different in both populations. In Japanese, a wild type of the Xba I 44 kb might have been derived from a Xba I 29 kb gene with a deletion of the restriction site on the 3' region of the CYP2D6 gene, and a defective gene of the Xba I 44 kb may have arisen in the wild type of the Xba I 44 kb allele.

Recently, Niznik et al. (1990) suggested the involvement of P-450dbl in the catabolism and processing of neurotransmitters such as dopamine subsequent to their reuptake into target cells in the brain. Cytochrome P-450dbl may play an important role not only in the liver but also in the brain, and a defective type of the Xba I 44 kb allele may occur more frequently in patients with IPD.

Acknowledgments

We are grateful to Dr. H. Shimabukuro, Okinawa Kyoudou Hospital, Dr. H. Higa, Naha City Hospital and Dr. M. Kuniyoshi, Okinawa Prefectural Hospital for collecting blood samples from patients with idiopathic Parkinson's disease.

References

Barbeau A, Cloutier T, Roy M, Plasse L, Paris S, Poirier J (1985) Ecogenetics of Parkinson's disease: 4-hydroxylation of debrisoquine. Lancet ii:1213–1215

Barbeau A, Roy M, Cloutier T, Plasse L, Paris S (1986) Environmental and genetic factors in the etiology of Parkinson's disease. Adv Neurol 45:299–306

Fonne-Pfister R, Meyer UA (1988) Xenobiotic and endobiotic inhibitors of cytochrome P-450dbl function, the target of the debrisoquine/sparteine type polymorphism. Biochem Pharmacol 37:3829–3835

Gonzalez FJ, Skoda RC, Kimura S, Umeno M, Zanger UM, Nebert DW, Gelboin HV, Hardwick JP, Meyer UA (1988) Characterization of the common genetic defect in humans deficient in debrisoquine metabolism. Nature 331:442–446

Heim M, Meyer UA (1990) Genotyping of poor metabolisers of debrisoquine by allele-specific PCR amplification. Lancet 336:529–532

Ishizaki T, Eichelbaum M, Horai Y, Hashimoto K, Chiba K, Dengler HJ (1987) Evidence for polymorphic oxidation of sparteine in Japanese subjects. Br J Clin Pharmacol 23:482–485

Kunkel LM, Smith KD, Bother SH, Borgaonkar DS, Wachtel SS, Miller OJ, Breg WR, Jones HW, Rary JM (1977) Analysis of human Y-chromosome-specific reiterated DNA in chromosome variants. Proc Natl Acad Sci USA 74:1245–1249

Langston JW, Ballard P, Tetrud JW, Irwin I (1983) Chronic parkinsonism in humans due to a product of meperidine-analogy synthesis. Science 219:979–980

Mahgoub A, Idle JR, Dring LG, Lancaster R, Smith RL (1977) Polymorphic hydroxylation of debrisoquine in man. Lancet ii:584–586

Nakamura K, Goto F, Ray WA, McAllester CB, Jacqz E, Wilkinson GR, Branch RA (1985) Interethnic differences in genetic polymorphism of debrisoquine and mephnytoin hydroxylation between Japanese and Caucasian populations. Clin Pharmacol Ther 38:402–408

Niznik HB, Tyndale RF, Sallee FR, Gonzalez FJ, Hardwick JP, Inaba T, Kalow W (1990) The dopanime transporter and cytochrome P450IID1 (debrisoquine 4-hydroxylase) in brain; resolution and identification of two distinct [^{3}H]-GBR-12935 binding proteins. Arch Biochem Biophys 276:424–432

Shahi GS, Moochhala SM, Das NP (1989) Depression of the hepatic cytochrome P450 monooxygenase system by the neurotoxin 1-methyl-4-phenyl-1,2,3,6-tetrahydropyridine (MPTP). Pharmacol Toxicol 64:107–110

Skoda RC, Gonzalez FJ, Demierre A, Meyer UA (1988) Two mutant alleles of the human cytochrome P-450dbl gene (P450C2D1) associated with genetically deficient metabolism of debrisoquine and other drugs. Proc Natl Acad Sci USA 85:5240–5243

Southern EM (1975) Detection of specific sequences among DNA fragments separated by gel electrophoresis. J Mol Biol 98:503–517

Yue QY, Bertilsson L, Dahl-Puustinen ML, Sawe J, Sjogvist F, Johansson I, Ingelman-Sundberg M (1989) Dissassociation between debrisoquine hydroxylation phenotype and genotype among Chinese. Lancet ii:870

Correspondence: Dr. I. Kondo, Departments of Human Ecology and Genetics, Faculty of Medicine, University of the Ryukyus, Okinawa 903–01, Japan.

Immunological aspects of neural transplantation

S. Kohsaka

Department of Neurochemistry, National Institute of Neuroscience, Kodaira, Tokyo, Japan

Summary

The mechanisms of the immunological rejection after xenogeneic neural transplantation were investigated. Tissue from a newborn mouse neocortical tissue was grafted into the third ventricle of a 4-week-old rat brain. The first event observed following the transplantation is the inosculation of blood vessels between those originating in the donor tissue and the host tissue. Thereafter, major histocompatibility complex (MHC) antigens were detected on the vascular endothelial cells at day 6, followed by the infiltration of cytotoxic/suppressor T-cells into the grafted tissue from day 8. These results raised the possibility that the expression of MHC antigens on the vascular endothelial cells renders the grafted tissues competent to initiate and participate in the immune reaction. This possibility was further confirmed by the finding that the host peripheral lymphocytes were sensitized by the donor MHC antigens, particularly class II, in the mixed lymphocyte culture experiment. Blood vessels in the grafted tissue are considered to be the initial target of the T-cells, because horseradish peroxidase perfusion experiment clearly demonstrated that the blood-brain barrier was destroyed at the 9th day after the transplantation.

Introduction

Neural transplantation is widely used as a viable approach to investigating the development of the nervous system. Moreover, the interest of this research tool is now focused on clinical application, particularly in association with Parkinson's disease.

It has long been believed that the central nervous system (CNS) is an immunologically privileged site because of the existence of a blood-brain barrier (Barker and Billingham, 1977), a poorly developed lymphatic system (Bradbury and Westrop, 1983) and lack of expression of major histocompatibility complex (MHC) antigens (Hart and Fabre, 1981). However, accumulating lines of evidence suggest that MHC-incompatible grafts are immunologically rejected in allogeneic (Mason et al., 1986; Nicholas et al., 1987a; Date et al., 1988) or xenogeneic (Inoue et al., 1985a,b; Brundin et al., 1985; Nakashima et al.,

the 1-4 th day
after transplantation
angiogenesis

the 5th day
inosculation of blood
vessels recirculation

the 6th day
expression of MHC antigens
on vascular endothelial cells

the 8th day
cell infiltration

the 14th day
destruction of blood vessels

after 14th day
rejection of the xenograft

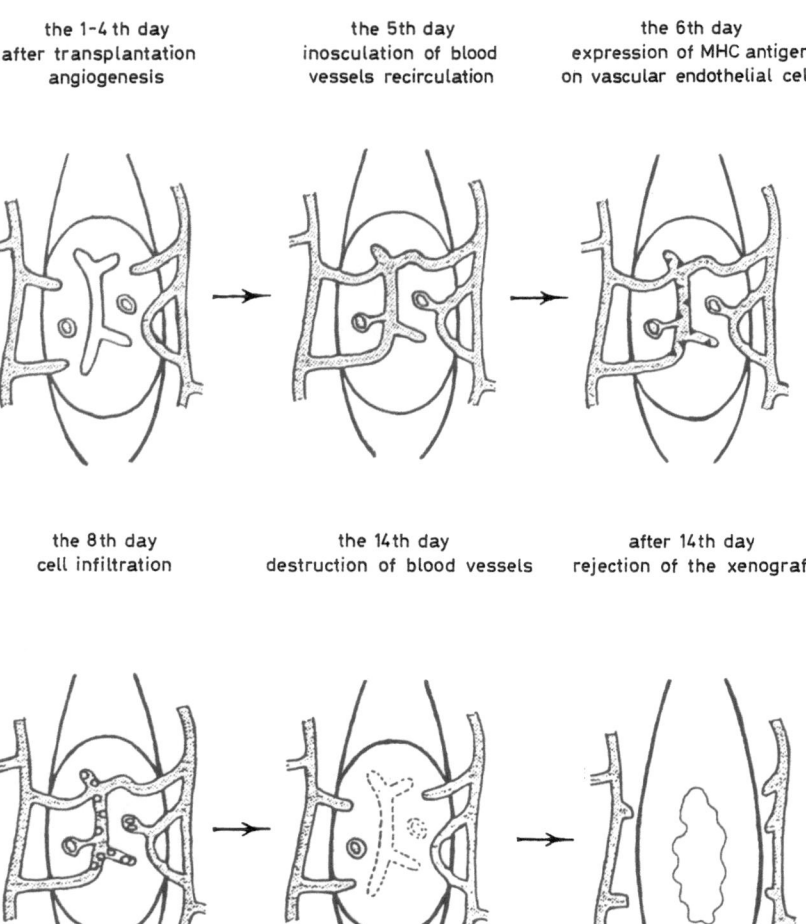

Fig. 1. Schematic illustrations of hypothesized process of immunological rejection in xenogeneic neural transplantation

1988) neural transplantation. We also demonstrated that the mouse neocortical tissue grafted into the third ventricle of rat brain was totally rejected by 4 weeks after transplantation, and that the rejection was prevented by administration of immunosuppressant cyclosporin A (Inoue et al., 1985a,b). However, the precise mechanism of the immunological rejection in neural transplantation is still unclear. This issue must be fully elucidated before the clinical application of intercerebral grafting. Previously, we examined the expression of Ia antigen,

T-cell infiltration and reconstruction of blood vessels in the mouse neocortical tissue grafted into the third ventricle of rat brain (Kohsaka et al., 1989; Nakano et al., 1989). The hypothesized mechanisms of the immunological rejection in xenogeneic neural transplantation based upon the results obtained in our previous experiments are illustrated in Fig. 1.

Cell infiltration

One to 3 days after transplantation, mild cell infiltration was detected in the grafted tissue. The infiltrating cells were mainly polymorphonuclear leukocytes. The leukocytes disappeared by day 4, and then mononuclear cells appeared from day 8 until the tissue was completely absorbed. Most of the mononuclear cells were immunohistochemically stained with OX-8 monoclonal antibody (Fig. 2), suggesting that the infiltrating cells are cytotoxic/suppressor T-cells.

Expression of Ia antigen

It is well documented that class II MHC antigen (Ia antigen) has a crucial role in immune responses after the transplantation of several organs such as

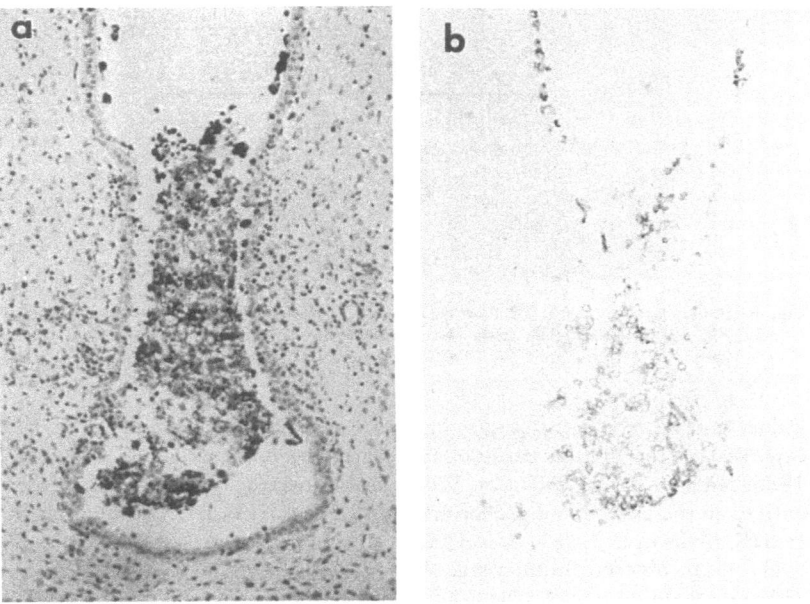

Fig. 2. Immunostaining with OX-8 monoclonal antibody. OX-8-positive mononuclear lymphocytes infiltrated into the grafted tissue. (a) Hematoxylin-eosin staining; (b) Immunostaining

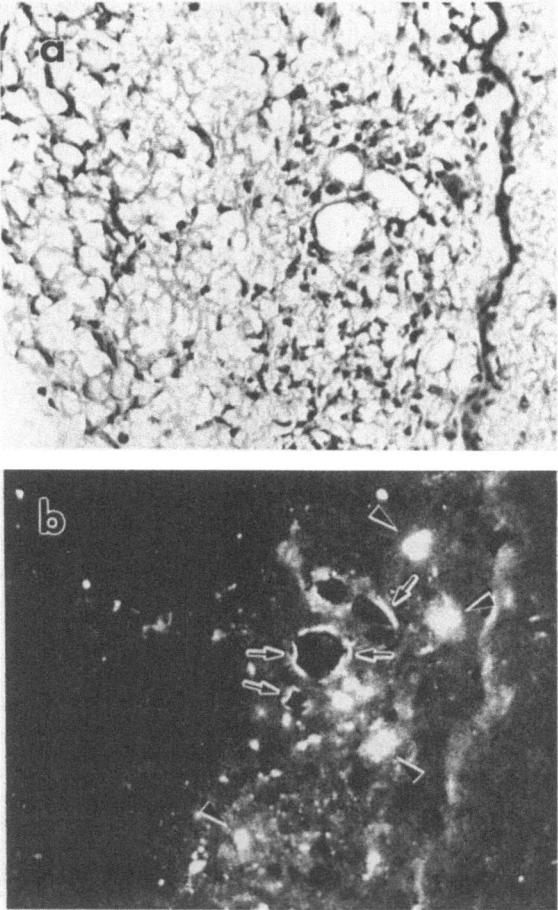

Fig. 3. Immunostaining with la monoclonal antibody. la antigens were expressed on vascular endothelial cells as indicated by arrows. (a) Hematoxylin-eosin staining; (b) Immunostaining

kidney and heart. Recent studies have demonstrated that la antigens are also expressed in the grafted tissue of the central nervous system (Mason et al., 1986; Nicholas et al., 1987a,b). We have also investigated the expression of la antigen in the grafted mouse neocortical tissue at various days after transplantation (Kohsaka et al., 1989). Until 5 days after transplantation, positive staining could not be observed in the tissue. It was first detected on day 6 in 1 out of 8 cases examined. On day 8, the la antigen was expressed in the grafted tissue in all cases. From immunohistochemical staining patterns, it was suggested that the la antigen was expressed on the vascular endothelial cells in the grafted tissue. The typical photomicrograph of the immunostaining with anti-mouse Ia antibody was shown in Fig. 3. In these experiments the fact that the vascular endothelial cells were stained with monoclonal antibody which specifically recognizes mouse Ia antigen leads to the interesting idea that the Ia-positive vascular endothelial cells originate in the donor tissue.

Reconstruction of blood vessels in the grafted tissue

We first examined the blood supply into the grafted tissue by India ink perfusion experiment. Invasion of India ink into the host blood vessels was first detected at the 5th day following the transplantation. The finding clearly indicated that the recirculation of the grafted tissue takes place at day 5, although blood vessels already exist from an earlier stage after the transplantation, suggested by the immunostaining pattern using anti-laminin antibody.

It is of interest to investigate whether the blood vessels in the grafted tissue originate in the donor tissue and receive a blood supply from the host brain. To answer this question, India ink perfusion experiments were performed in combination with immunohistochemical staining using monoclonal antibody which specifically recognizes mouse endothelial surface antigen (MESA-1) (Ghandour et al., 1982). The result clearly indicated that the blood was supplied to the blood vessels originating in the mouse neocortical tissue (Fig. 4). Furthermore, the result leads to the possibility that blood vessels originating in the donor and the host brain inosculate with each other about 5 days following transplantation. To confirm this possibility, electron microscopic immunohistochemistry was performed using MESA-1 antibody. As shown in

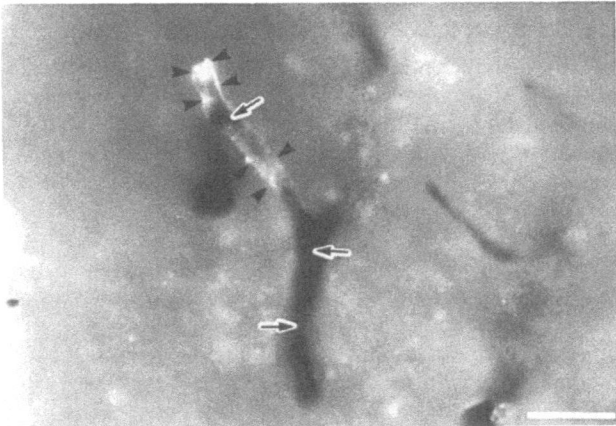

Fig. 4. Immunostaining with MESA-1 monoclonal antibody following India ink perfusion. Blood vessels in the grafted tissue were stained by India ink (arrows) and a part of the vessel was clearly stained by MESA-1 antibody (arrowheads)

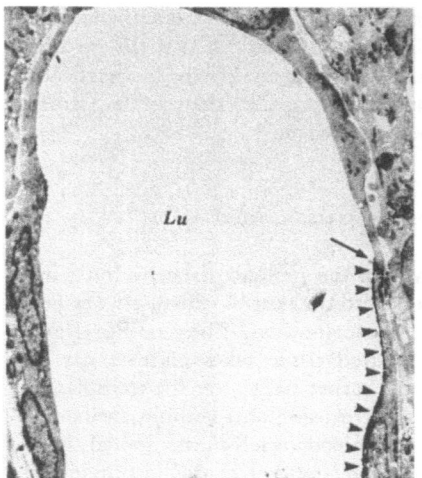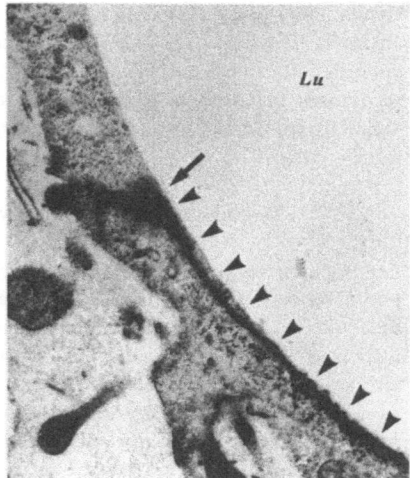

Fig. 5. Electron microscopic immunohistochemical staining of vascular endothelial cells with MESA-1 monoclonal antibody. Arrowheads indicate the deposits of the reaction products. Arrows indicate the interface between immuno-positive and immuno-negative endothelial cells

Fig. 5, the blood vessels were chimeric, consisting of two different kinds of vascular endothelial cells in terms of immunostaining patterns. This mosaic reconstruction of blood vessels strongly supports the idea that the blood vessels originating in the mouse brain inosculate with those of the host rat brain.

From these results, we have speculated the possible mechanisms for the immunological rejection in xenogeneic neural transplantation as follows: 1) the blood vessels originating in the donor tissue undergo reconstruction until 4 days after transplantation; 2) the blood vessels originating in the donor and the host tissue inosculate with each other at day 5; 3) MHC xenoantigen is expressed on the vascular endothelial cells originating in the donor tissue at day 6; 4) T-cells began to infiltrate into the donor tissue from day 8; 5) the donor tissue-originated blood vessels were attacked by the T-cells and the blood circulation ceased from about 14 days after the transplantation; and 6) donor tissue degenerates possibly because of ischemic condition.

Sensitization of the host peripheral lymphocytes

To confirm the hypothesis of the rejection mechanisms described above, we further examined the following items. 1) Are host peripheral lymphocytes sensitized by the mouse MHC antigen expressed on the vascular endothelial cells? 2) Are the blood vessels in the donor tissue attacked by T-cells as the first target during the course of the immunological rejection?

Nicholas et al. (1987b) recently reported that the donor tissue was rejected in allogeneic neural transplantation. They examined the systemic sensitization

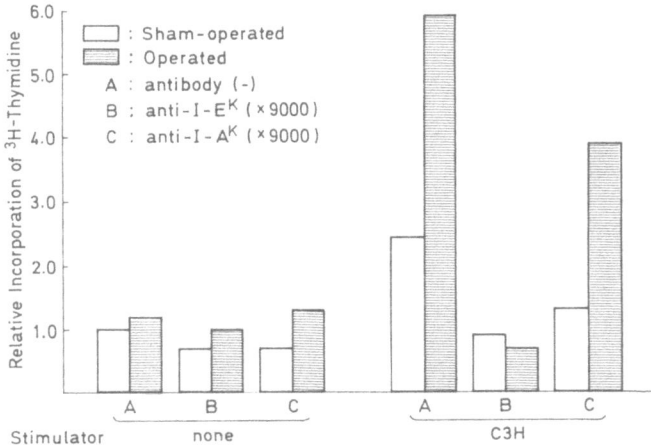

Fig. 6. Proliferative responses of host peripheral lymphocytes in mixed lymphocyte culture. The host lymphocytes as responder cells were prepared at the 14th day after transplantation. C3H mouse splenocytes were used as stimulator cells

of the peripheral lymphocytes by H-2 alloantigens by using mixed lymphocyte cultures. They concluded that intraventricular neural transplants, while recognized and affected by cells of the host immune system, do not elicit a detectable systemic sensitization to class I H-2 alloantigens. They speculated that rejection of neural transplants may depend on sensitization to class II H-2 alloantigens. Therefore, we have investigated the systemic sensitization of the host peripheral lymphocyte to class II MHC antigens (I-A^K and I-E^K). Newborn mouse (C3H) neocortical tissue was transplanted into the third ventricle of rat brain according to the same method described previously (Kohsaka et al., 1989). At the 14th day after the transplantation, the peripheral lymphocytes were prepared from the axial lymph node of the host as a responder. As a stimulator, normal splenocytes were obtained from adult C3H mouse spleen. Mixed suspensions of the splenocytes and host peripheral lymphocytes were cultured for 3 days and the sensitization by class II MHC antigens was examined by counting the incorporated ^{3}H-thymidine in the lymphocytes. As shown in Fig. 5, the host lymphocytes significantly responded to the splenocytes of C3H mouse and this response was significantly inhibited by anti-I-E^k monoclonal antibody, while the response was suppressed only to a small extent by anti-I-A^k monoclonal antibody. These results suggest that the host peripheral lymphocytes were mainly sensitized by I-E antigen among class II MHC products.

Destruction of blood-brain barrier (BBB) in the grafted tissue

In order to determine whether the blood vessels in the grafted tissue are attacked by activated T-cells, we have examined the existence of a blood-brain

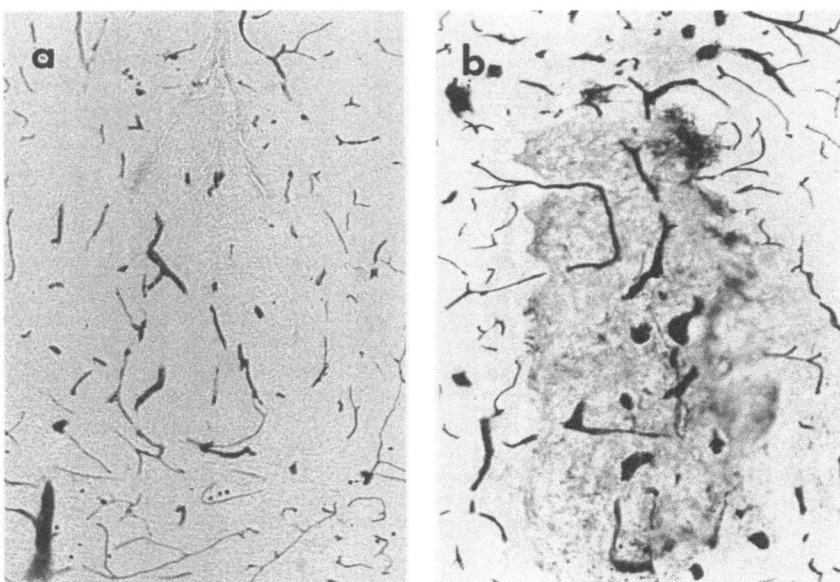

Fig. 7. Leakage of HRP from the blood vessels in syngeneic (a) and xenogeneic (b) grafts tested on the 9th day after the transplantation. HRP was leaked in the xenograft (b) but not in the syngraft (a)

barrier (BBB) by using the horseradish peroxidase (HRP) perfusion method (Kohsaka et al., 1989; Rosenstein, 1987). At the 9th day after the transplantation, HRP was injected from the host blood vessel, and leakage of the injected HRP was examined in the donor tissue. In this experiment, newborn rat neocortical tissue was also implanted into rat brain (syngeneic neural transplantation) and the leakage of the injected HRP into the donor tissue was also examined as a control. Figure 7 shows typical photomicrographs of the HRP stainings of the sections from the control and the experimental animals. HRP did not leak into the donor tissue in syngeneic neural transplantation (Fig. 7a). In contrast, a large amount of leaked HRP was detected in the grafted mouse neocortical tissue in the xenogeneic neural transplantation (Fig. 7b). The results obtained from the experiments clearly demonstrated that the BBB of the blood vessels in the donor tissue is damaged possibly because of the sensitized T-cells.

Acknowledgments

The present studies were supported by a Grant-in-Aid for Scientific Research on Priority Area from the Ministry of Education, Science and Culture, Japan. The author is grateful to his colleagues, Drs. T. Nakano, T. Shinozaki and S. Toya in the Department of Neurosurgery, Keio University School of Medicine and to Dr. K. Takei in the

Department of Physiology, Keio University School of Medicine. The author also thanks Ms. M. Kamei for the preparation of this manuscript.

References

Barker CF, Billingham RF (1977) Immunologically privileged sites. Adv Immunol 25:1–54

Bradbury MWB, Westrop RJ (1983) Factors influencing exit of substances from cerebrospinal fluid into deep cervical lymph of the rabbit. J Physiol 339:519–534

Brundin P, Nilsson OG, Gage FH, Bjorklund A (1985) Cyclosporin A increases survival of cross-species intrastriatal grafts of embryonic dopamine-containing neurons. Exp Brain Res 60:204–208

Date I, Kawamura K, Nakashima H (1988) Histological signs of immune reactions against allogeneic solid fetal neural grafts in the mouse cerebellum depend on the MHC locus. Exp Brain Res 73:15–22

Ghandour S, Langley K, Gombos G, Hirn M, Hirsch MR, Goridis C (1982) A surface marker for murine vascular endothelial cells defined by monoclonal antibody. J Histochem Cytochem 30:165–170

Hart DNJ, Fabre JW (1981) Demonstration and characterization of Ia-positive dendritic cells in the interstitial connective tissue of rat heart and other tissues, but not brain. J Exp Med 154:347–361

Inoue H, Kohsaka S, Yoshida K, Otani M, Toya S, Tsukada Y (1985a) Cyclosporin A enhances the survivability of mouse cerebral cortex grafted into the third ventricle of rat brain. Neurosci Lett 54:85–90

Inoue H, Kohsaka S, Yoshida K, Otani M, Toya S, Tsukada Y (1985b) Immunohistochemical studies on mouse cerebral cortex grafted into the third ventricle of rats treated with cyclosporin A. Neurosci Lett 57:289–294

Kohsaka S, Shinozaki T, Nakano Y, Takei K, Toya S, Tsukada Y (1989) Expression of Ia antigen on vascular endothelial cells in mouse cerebral tissue grafted into the third ventricle of rat brain. Brain Res 484:340–347

Mason DW, Charlton HM, Jones AJ, Lavy CB, Puklavec M, Simmonds SJ (1986) The fate of allogeneic and xenogeneic neural tissue transplanted into the third ventricle of rodents. Neuroscience 19:685–694

Nakano Y, Takei K, Toya S, Tsukada Y, Ghandour S, Kohsaka S (1989) Mosaic reconstruction of blood vessels in mouse neocortical tissue transplanted into the third ventricle of rat brain. Brain Res 496:336–340

Nakashima H, Kawamura K, Date I (1988) Immunological reaction and blood-brain barrier in mouse-to-rat cross-species neural graft. Brain Res 475:232–243

Nicholas MK, Antel JP, Stefansson K, Arnason BGW (1987a) Rejection of fetal neo-cortical neural transplants by H-2 incompatible mice. J Immunol 139:2275–2283

Nicholas MK, Steffansson K, Antel JP, Arnason BGW (1987b) An in vivo and in vitro analysis of systemic immune function in mice with histologic evidence of neural transplant rejection. J Neurosci Res 18:245–257

Rosenstein JM (1987) Neocortical transplants in the mammalian brain lack a blood-brain barrier to macromolecules. Science 235:772–774

Correspondence: S. Kohsaka, Department of Neurochemistry, National Institute of Neuroscience, 4-1-1 Ogawa-higashi, Kodaira, Tokyo 187, Japan.

Epidemiology of Parkinson's disease-interplaying genes and environmental factors

K. Kondo

Department of Public Health, Hokkaido University, School of Medicine, Sapporo, Japan

Summary

Epidemiology is one of the methods for solving a complex interplay of genes and environmental factors in Parkinson's disease (PD).

A multifactorial inheritance has been proposed. Low concordance in twins refuted the role of genes in PD, but the small sample size did not allow a final conclusion. This author pooled twin and sib data from various sources and obtained 65% as the heritability of PD.

Risk factors are of importance in resolving the interplay. Well-accepted factors are age, smoking, preclinical personality profiles, and affection of the relatives with PD. We evaluated 271 questions in 166 cases along with two each of controls. PD-prone lifestyles were: 1) non-Japanese diet; 2) less smoking 3) less drinking; 4) fewer hobbies and if any, indoor hobbies requiring no partner; 5) interested only in business, narrow surroundings, etc.; 6) more habitual constipation, less circulatory disease; 7) less physical exercise; and 8) preclinical personality.

PD is probably a consequence of a long-standing age-dependent process modified by the interaction of a genetic predisposition and cumulated effects of numerous neurotoxins exposed for decades of time.

Introduction

Rational therapeutics based on catecholamine metabolism is now followed by studies of the mechanism of neuronal death in Parkinson's disease (PD), motivated by reports that 1-methyl-4-phenyl-1,2,3,6 tetrahydropyridine (MPTP) caused PD-like syndromes in man and animals.

Epidemiology describes the frequency and its determinants of a disease, discloses inherited or environmental risk factors, and when possible, prevents the disease by introducing intervention programs.

Even in diseases influenced by numerous genes and environmental factors, each having relatively minor effects, well-planned biometrical studies can often elucidate how these factors interplay to cause the disease.

After reviewing descriptive-epidemiological data as well as results of biometrical-genetic analyses, this report discusses risk factors, especially life-styles in PD, with some emphases on the Japanese data as a baseline for better studies to solve the complex interplay of factors underlying PD.

Frequency and its determinants

The major results are summarized in Table 1. Frequency is given in terms of the point prevalence rate, or the number of patients per 100,000 surviving at a given time in a defined population. The values range between 60–187 for Caucasians with an average of 117.6 based on 16 reported figures before 1984, but 34.3–55.0 for 4 areas in Japan averaging 41.1, and 57 for a pooled population of 6 cities in China (Li et al., 1985). There is only one reliable Black American data set indicating that the Black rate was about half of the Caucasian rate in Mississippi, U.S.A. (Schoenberg et al., 1980). While the figures showed

Table 1. Parkinson's disease in the Western Countries and Japan

	Western	Japan
A. Clinical-pathological:		
Clinical features	Triad	Same, less requirement of l-dopa
Pathological features	Degeneration of nigra, etc	Same
B. Epidemiological patterns:		
Sex difference	None	None
Prevalence rate	$100 - 150 \times 10^{-5}$	$40 - 50 \times 10^{-5}$
Secular trend	None	None
Arenal difference	None ?	None
Incidence rate	Increase with age	Same
Associations with;		
Other diseases	?	None
HLA, etc	None	None
Smoking	Reverse	Reverse
Drinking	?	Reverse
Preclinical features;		
Personality	Rigid	Rigid
Constipation	Yes	Yes
Lifestyle	?	Unique
Physical exercise	?	Less
C. Genetic patterns:		
Single-gene models	Rejected	Rejected
Multifactorial model	Accepted	Accepted
Heritability	65%	60%
Twin studies	Low concordance	?
Chromosome study	Normal	?

a difference between the Caucasians and the Orientals, they are relatively stable in a given race.

No established evidence supported the differences of the frequency between the sexes, over time, among the areas despite climatic, latitudinal or socio-economic diversities. Unlike motor neuron diseases, no established high-risk area or areal clusterings are known in PD.

Age-specific incidence rates are available for three areas. In all, the rate is very low before the age of 40, but rises sharply between the ages of 40 and 80, seemingly declining thereafter. Epidemiologists assume that such a decline is due to insufficient case ascertainments in the elderly, and the rate is, in fact, monotonously rising.

Risk factors and case-control studies

Factors predating clinical onset of a disease are called risk factors to the disease when individuals with the factors have an elevated probability of being affected. Risk factors are defined statistically and are not necessarily the cause of disease. In obscure disease, they are useful mainly in three ways: 1) as clues for the cause, 2) to identify high-risk individuals, and 3) when eliminated, to prevent the disease.

Any statistically adequate method can be used to identify a risk factor. The following methods have been applied to PD; a) associations with other diseases in clinical and autopsy series, b) associations with inherited markers, c) seroepidemiological studies for possible viral etiologies, and d) case-control studies for histories prior to onset (Kondo, 1986). Methods a, b, and c disclosed parctically negative findings. Few pioneering case-control studies evaluated medical histories and some habits like smoking, revealing that circulatory diseases and smoking histories are less.

Epidemiologists in Japan are particularly interested in lifestyles as risk factors to obscure chronic disorders in the hope of obtaining a clue to the causes and as a means of their prevention through advice on a healthier life.

A multicenter case-control study of PD was introduced in the early 1980s. This study used 166 cases and one each of like-sexed and like-aged ($\pm$ 5 years) hospital and community controls, asking 271 questions.

In this first attempt to evaluate PD-prone lifestyle profiles, the risk factors were 1) non-Japanese diet, 2) less smoking, 3) less drinking, 4) fewer hobbies and if any, indoor nonsporting hobbies requiring no partner, 5) narrow interests, only in business, own surroundings, limited human relations etc., 6) more habitual constipation, 7) less circulatory disease, 8) less physical exercise, 9) rigid preclinical personality (Kondo, 1986).

Reviewing available data including ours, to date, we found that the established risk factors to PD are advanced age, less smoking, rigid preclinical personality and affection of the relatives. Furthermore, possible risk factors include head injuries, low blood pressure, habitual constipation, reduced physical activity, etc.

Despite some methodological inadequacies, a case-control study is very useful for a quick screening of candidate risk factors. It is also useful to test hypotheses proposed from other areas of investigation. Such studies will be particularly rewarding, moreover, if good hypotheses are produced to be tested biometrically, based on the studies in various other disciplines of study.

Genetic studies

About 5% of PD patients have a "positive family history".

1. Genetic analyses

The mode of transmission was assumed to be autosomal-dominant in early reports of families with PD, without analyzing the family data.

In formal-genetic analyses, observed family patterns are rigorously compared with the expected patterns under a few possible modes of transmission. While hypotheses that show inconsistencies are rejected, those showing that the expected and the observed patterns are identical are retained. The author reported this kind of study, and while rejecting all single-gene hypotheses, provided reasonable evidence for a multifactorial inheritance with a "threshold effect" (Table 2). Namely, PD clinically manifests when a "liability" is above a certain level, in which the value of such liability is influenced by numerous genes and environmental factors (Kondo et al., 1973).

Table 2. Expected and observed family patterns, various genetic hypotheses, Parkinson's disease

Expected patterns	Compatibility with the observed patterns
A. *Single-gene inheritances*:	
X-linked recessive. Parents normal; 1/2 sons but no daughter affected; no paternal males affected	no
X-linked dominant. All daughters but no son of affected fathers affected	no
autosomal recessive. No parent affected; 1/4 children affected	no
autosomal dominant. One parent affected; 1/2 children affected; few sporadic mutants	no
B. *Multifactorial inheritance with a threshold effect*:	
sib recurrence rate is square root of prevalence rate	yes
risk decreases in remote relatives	yes
concordance rate larger in monozygotic than dizygotic twins	yes?
risk increases with younger age at onset	yes
risk increases with more sibmates already affected	yes

Adapted from Kondo et al. (1973) and other reports

How a genetic variability contributes to the overall variability of liability is given by estimating the "heritability". Based on Falconer's model (1965), the value was 80% for the data collected in Minnesota, USA (Kondo et al., 1973), but a new estimate was obtained by pooled twin data as explained later.

2. Concordance in the twins with Parkinson's disease

Monozygotic twins (MZ) are genetically identical and any intrapair difference is due to environmental reason(s). Dizygotic twins (DZ) are genetically as close as siblings, but the partners share intrauterine environments.

In theory, twins are useful for evaluating the nature/nurture problem in a given trait, but in reality, the results are often perplexing, mainly because of difficulties of collection of sufficient data without biases or errors.

A relatively low concordance rate in MZ's with PD was considered to underestimate the role of genes (Ward et al., 1983). The essential point with the twin data is to compare groups of individuals who share genes to known extents. The same purpose as intrapair comparisons in MZ vs DZ, however, can also be achieved based on other types of populations which are larger and more accessible in epidemiological surveys. Table 3 shows that frequency of PD grows as the extent to which the members share genes becomes larger, apparently indicating the role of genetic factors in PD.

Based on the Smith method (1970), the heritability was about 65%, calculated from this data.

3. Lessons from biometrical genetics

Contributions and limitations of a multifactorial theory of PD are that 1) liability is hypothetical and directly unmeasurable, 2) genes and environmental

Table 3. Frequencies of Parkinson's disease in some Caucasian populations sharing genes to various extents (Kondo, 1989)

Populations	% Genes shared	Frequencies (%)
General populations	Nearly zeno	less than 1
Siblings	50.0	5.7
Parents	50.0	5.7
Twins		6.4
Dizygotic twins	50.0	6.7
Monozygotic twins	100.0	6.3

These are pooled data of those aged over 60 in the most of the quoted reports. Frequency is expressed as the prevalence rate for the general populations, as the recurrence rate for siblings and parents, and as the concordance rate for twins

factors are not identified individually, 3) single-gene heredities were rejected, 4) heritability suggested that the relative role of genes is fairly large, 5) the theory predicted that most cases, especially elderly cases, are sporadic, and familial aggregation is likely to occur in cases manifesting at younger age, patterns of the familial aggregation being highly variable, with some families showing two or three cases while others are reminiscent of dominant inheritance, and 6) a model was theoretically forwarded that PD is like the tip of an iceberg, visible in those who have a liability elevated above the threshold, either due to unfavorable genes or to unfavorable environmental to which factors they were exposed, or combinations of the two.

Gene-environment interplay

Interplay of the factors in multifactorial inheritance can usually only be solved by evaluating each candidate factor with rigorous methods. For the detection of environmental factors and some host-related factors, a case-control study still remains the only feasible approach, whereas molecular and bio-chemical genetics also provide useful methods of analyses for this purpose. These include studies of inherited markers such as HLA, polymorphisms of enzymes influencing metabolic rates of various substances, exogenous and endogenous, which affect pathomechanisms underlying PD. This highly promising area has just opened (Grima et al., 1987; Kondo and Kanazawa, 1991).

1. Complex situations where factors are difficult to identify individually

Causal relationship in a given disease can easily be detected when a short-time exposure to a single, strong and specific agent rapidly causes a specific, easily recognizable syndrome. Acute MPTP experiments represent such a situation.

The action of the environmental factors in PD is probably not so simple, however, and it may be that cumulative effects occur after a long-time exposure to small amounts of various agents interacting with genetic predispositions influenced by plural loci, and superimposing themselves on the aging process. The agents may be plural, each within unharmful amounts alone but jointly, they become substantially harmful. These situations are proven in experimental carcinogenesis, and unlike some occupational cancers due to a heavy short-term exposure to a single or a few identifiable carcinogens, most cancers are caused in these ways.

An additional difficulty in the etiological study of PD is that its clinical onset is preceded by a long preclinical period (Table 4).

This is a very unusual situation rarely encountered in conventional epidemi-ological studies. Whereas a geneticist is primarily interested in understanding the mechanisms of familial aggregation, an epidemiologist wishes to identify "association" of a given disease with a given factor in question, not necessarily

Table 4. Evidences supporting that Parkinson's disease is a final manifestation of a long subclinical course

A. *Clinical features*:
Subtle subjective difficulties often precede objective signs; symptoms sometimes intermittent initially; some patients manifest PD after a triggering event.

B. *Preclinical features*:
Unique personality profile before onset; abstinence from smoking and drinking since young ages, depressed state before onset.

C. *Familial aggregation*:
Multifactional genetic predisposition with a high heritability (Kondo et al., 1973).

D. *Drug-induced parkinsonism*:
That this is more frequent in advanced ages may indicate increasing loss of nigral neurons predispose this complication; drug-induced patients have more relatives of idiopathic PD (Myrianthopoulos et al., 1969).

E. *Metabolic features*:
Patients detoxify debrisoquine slowly (Barbeau et al., 1985).

F. *Neuroimaging*:
Seemingly normal indivuduals exposed to MPTP showed reduced activity on fluoro-dopa PET scanning(Calne et al., 1985); patients with Guamanian ALS without PD showed the same finding.

analyzing the mechanism. Ottman (1990) proposed a unique method of approach to gene-environment interplay of a given disease without involving knowledge of its pathogenesis (Table 5). In PD, such analyses have not been made but may give useful clues.

Table 5. Expected patterns of relative risks for five models of gene-environment interactions, by family history and the putative risk factors (Ottman, 1990)

Risk factor	Familial		Sporadic	
	Yes	No	Yes	No*
Model A	+	1	+	1
B	+ +	1	+	1
C	+ +	+	1	1
D	+	1	1	1
E	?	+	+	1

* Reference groups with assigned relative risk 1, + elevated relative risk, + + highly elevated relative risk compared with the reference group.

Model A. The genotype (G) increases expression of the risk factor(R) (Example; Phenylketonuria and high phenylalanine to cause mental retardation)

Model B. G exacerbates the effect of R (Xeroderma pigmentosum and ultraviolet radiation to skin cancer)

Model C. R exacerbates the effect of G (Barbiturate and porphyria variegata to skin disorder)

Model D. G and R required to raise the risk (G6PD deficiency and fava bean to hemolytic anemia)

Model E. G and R each influence by themselves (alpha-1 antitrypsin deficiency and smoking to pulmonary emphysema)

2. Significance of lifestyles in adult diseases

This cannot be overemphasized in most chronic multifactorial diseases occurring in the adult to elderly years; e.g., hypertension, hyperlipidemias, arteriosclerosis, hyperuricemia, etc., as well as accidents secondary to these disorders such as stroke or heart attack. In these diseases, specific inherited abnormalities sometimes increase the risk. For example, hypercholesterolemias due to the inherited defects of the LDL receptors predispose to premature atherosclerosis and myocardial infarction. Even in these cases with a specific inborn defect, however, lifestyles over decades, including diet and sporting activity, strongly influence the course of the arterial disease and its complications.

Neurons are postmitotic cells. They die one by one, but owing to numerical redundancies and compensatory mechanisms, the functional reserve of a given neuronal system is very large so that a functional impairment does not manifest within the usual lifespan of man. ɟ

It is postulated that cumulated effects of numerous environmental factors each having minor but adverse influences may promote senescence and death of the neurons in a system in susceptible individuals. It is suggested that such a process takes a prolonged period of years. Riederer and Wuketich (1976) estimated the reduction rate of the caudate dopamine to be 12.8% per decade of life in the normals but about 29-47% in PD, and that the process started 20–30 years before clinical onset. This estimation is compatible with evidence shown in Table 4, and it is extremely important to elucidate factors that facilitate this process.

Circumstantial evidence is adequate, but we still need more direct information about inherited variants as well as postnatal factors, including lifestyles, that facilitate neuronal death and are thus associated with elevated risks of PD.

Food preferences are particularly important as putative lifestyle factors, as preliminarily suggested in our case-control study. At least in acute experiments in squirrel monkeys, MPTP-resembling tetrahydroisoquinoline produced a condition comparable to human PD (Yoshida et al., 1991). Quoting and amplifying the data on amounts of MPTP-related substances in various foods (Niwa et al., 1989), one should be able to estimate lifelong exposures to such substances in cases and in control groups. A dose-response relationship may be analyzed to establish whether PD is associated with increased exposures to such substances or other food constituents.

Exercise and sporting are also important in evaluateing whether some peculiarities of this aspect are associated with subsequent PD. In food-associated diseases, it is usual that exercise/sporting is also associated. In our stated case-control study, those who later showed PD were less physically active in all decades prior to onset starting from the juvenile age. It is unknown whether such a pattern is a preclinical manifestation of PD.

Psychosocial activities are known to influence Alzheimer's disease (Kondo and Yamashita, 1990). Their significance in PD awaits an extensive evaluation, as it is likely that certain personality types are prone to this disease.

International comparison is a useful method to analyze a difficult problem like the one we deal with. Comparison of the risk factors in the countries showing a different frequency of the disease concerned may give a breakthrough for better understanding of this nature/nurture problem. For example, in the following three groups,

1. Japanese in Japan showing a prevalence rate of PD, 50−60 per 100,000;
2. Japanese-Americans born in the U.S.A. to purely Japanese parents; and
3. Caucasians born in the U.S.A. or Europe, with the rate of PD, 100−150 per 100,000;

the second group involves those who are genetically identical to the first group, but subject to almost the same environment as the third group. The frequency of PD may show a gradient among these three groups, and a new case-control study utilizing the cases as index cases from these groups may give explanations for such a gradient and significances of the factors, especially lifestyles.

Acknowledgments

Cordial thanks are due to Ms. C. Onishi for her excellent assistance. Supported by grants from the Neurodegenerative Diseases Research Committee, the Japan Ministry of Health and Welfare.

References

Barbeau A, Cloutier T, Roy M, Plasse L, Paris S, Poirier J (1985) Ecogenetics of Parkinson's disease; 4-hydroxylation of debrisoquine. Lancet ii:1213−1216

Calne DB, Langston JW, Martin WR, Sloessl AJ, Ruth TJ, Adam MJ, Pate BD, Schulzer M (1985) Positron emission tomography after MPTP; observation relating to the cause of Parkinson's disease. Nature 317:246−248

Falconer DS (1965) The inheritance of liability to certain diseases estimated from the incidence among relatives. Ann Hum Genet 29:51−76

Grima B, Lamouroux A, Boni C, Julien J-F, Javoy-Agid F, Mallet J (1987) A single human gene coding multiple tyrosine hydroxylases with different predicted functional characteristics. Nature 326:707−711

Kondo K (1986) Epidemiological evaluation of risk factors in Parkinson's disease. In: Yahr MD, Bergmann KJ (eds) Parkinson's disease. Raven Press, New York, pp 289−293

Kondo K (1989) Interplay of the environment and genes in Parkinson's disease. Adv Neurol Sci (Jpn) 33:759−765

Kondo K, Yamashita (1990) A case-control study of Alzheimer's disease in Japan; association with inactive phychosocial behaviors. In: Hasegawa K, Homma A (eds) Psychogeriatrics: biomedical and social advances. Excerpta Medica, Amsterdam, pp 49−53

Kondo K, Kurland LT, Schull WJ (1973) Parkinson's disease, genetic analysis and evidence of a multifactorial etiology. Mayo Clin Proc 48:465−475

Kondo I, Kanazawa I (1991) Association of Xba I allele (Xba I 44 kb) of the human cytochrome P-450 dbI (CYP2D6) gene in Japanese patients with idiopathic Parkinson's disease (this volume)

Li SC, Schoenberg BS, Wang C, Cheng X, Rui D, Bolis CL, Schoenberg DG (1985) A prevalence survey of Parkinson's disease and other movement disorders in the People's Republic of China. Arch Neurol 42:655−657

Myrianthopoulos NC, Waldrop IN, Vincent BL (1966) A repeated study of hereditary predisposition in drug-induced parkinsonism. In: Barbeau A, Brunette JB(eds) Progress in neurogenetics. Excerpta Medica, Amsterdam, pp 486–491

Niwa T, Yoshizumi H, Tatematsu A, Matsuura S, Nagatsu T (1989) Presence of tetrahydroisoquinoline, a parkinsonism-related componend in foods. J Chromatogr 493:347–352

Ottman R (1990) An epidemiologic approach to gene-environment interaction. Gen Epidemiol 7:177–185

Riederer P, Wuketich S (1976) Time course of nigrostriatal degeneration in Parkinson's disease. J Neural Transm 38:277–301

Schoenberg BS, Anderson DW, Haerer AF (1985) Prevalence of Parkinson's disease in the biracial population of Copiah Country, Mississippi. Neurology 35:841–845

Smith C (1970) Heritability of liability and concordance in monozygotic twins. Ann Hum Genet 34:85–91.

Ward CD, Duvoisin RC, Ince SE, Nutt JD, Eldridge R, Calne DB (1983) Parkinson's disease in 65 pairs of twins and in a set of quadruplets. Neurology 33:815–824

Yoshida M, Ogawa M, Nagatsu T (1991) Can tetrahydroisoquinoline (TiQ) or N-methyl-TiQ produce parkinsonism? (this volume)

Correspondence: Dr. K. Kondo, Department of Public Health, Hokkaide University, School of Medicine, N 15 W7, Kitaku, Sapporo, Japan 060.

Topics of special interest in research of Parkinson's disease

H. Narabayashi

Prof. Emeritus, Juntendo University, and Neurological Clinic, Tokyo

Parkinson's disease (PD) has been recognized as a idiopathic, degenerating disease presenting a mixture of a variety of different grades of rigidity, tremor, and akinesia. Today it is established that the axial pathology of the disease is the nigrostriatal DA deficiency, for which the treatment using levodopa or DA agonists produces markedly favourable results. Due to advances in drug treatment and resulting prolongation of the life span of PD patients, symptoms that had previously been not considered as being of core importance, have slowly become recognized as being important. These symptoms and difficulties in the later stage of the patients under long-term levodopa treatment for more than several years are disturbances of postural control and gait, emotional deterioration, and deterioration of cognitive function.

For understanding and treatment of a chronic and progressive disease like PD presenting a variety of symptoms, two investigative approaches are considered important. One is to elucidate the intracerebral pathophysiological mechanism underlying each of these symptoms such as rigidity, tremor, or akinesia. This avenue, which covers the anatomical, physiological, and biochemical investigations firstly based on neuropathological knowledge, is the classical one but is also today's interest in the dynamic network within the basal ganglia. This will be described later by the author. The second approach, which has rapidly developed in recent years, is the investigation into the etiopathogenesis of the degenerative process specific for PD, i.e. into the nature of the slowly progressive neuronal death. This field may include the molecular biological investigation of the mechanism of more general cell death and genetic analysis.

§I. Back in May 1952 the author started the analytical approach by initiating stereotaxic pallidotomy for treatment of the rigidity and tremor of the disease, which surgery was later switched to thalamotomy. Microelectrode recording technique during surgical procedure was introduced in the author's operation theater as a routine method in 1972, and since then it has become possible to interpret the surgery-elicited changes in clinical symptoms in exact anatomical and physiological terms. Clinical effects on rigidity and tremor, the underlying pathophysiological mechanisms of these symptoms, and observations on the long-term postoperative follow-up study are described in the chapter, "Role of stereotaxic surgery in treatment of Parkinson's disease" in this book and in several other papers by the author (Narabayashi, 1990a).

Although the pace-maker of the tremor-generating mechanism is still not fully explained, pathological tremor, either parkinsonian or non-parkinsonian, is now almost completely alleviated by stereotaxic surgery on the Vim (ventral intermediate nucleus) of the thalamus without side effect. Tremor is interpreted as a circuit phenomenon based on both central and peripheral mechanisms with the Vim being a structure of key importance.

Rigidity is alleviated by lesioning of the VL (ventrolateral nucleus), the nucleus that lies just anterior to the Vim in the base of the thalamus and receives pallidal afferents. The pallido-VL-thalamic pathway is now considered responsible for producing rigidity and lies under the control of the nigrostriatal DA system.

§II. Akinesia is considered the most important core symptom of the disease, although the term itself has not been clearly defined and is used in a relatively vague manner. Akinesia is also analysed in the similar way and is classified into three subgroups. Type-I akinesia is slowness and unskillfulness of movement and is secondary to muscle rigidity, for it is almost completely abolished when rigidity has been removed by thalamic surgery. Type-II akinesia is also called primary askinesia; it is not related to rigidity but responds well to levodopa. Type III-akinesia seems more non-specific and is seen mostly in the chronic stage of the disease under long-term levodopa treatment, the symptoms of which have briefly been mentioned at the beginning of this chapter. Within the Type-III akinesia category, the symptom of "freezing" in gait and in repetitive movement is interested. It is frequently seen in longstanding PD cases and is difficult to control by levodopa. The underlying mechanism of gait-freezing was analysed by *Imai*, as will be described later in this chapter, by studying cases of "pure akinesia", which is the condition presenting the kinésie paradoxale type gait-freezing without any sign of rigidity and tremor and not responding to levodopa at all. Difficulty of rhythm formation over 2.5 hertz in repetitive movement in these patients was analysed as one of the underlying phenomena of freezing by Nakamura and Narabayashi (1976).

Gait freezing in the later stage of PD patients was hypothesized to be related to a central NE deficiency after the confirmation of lowered NE metabolism in the brains and CSF in the longstanding cases. L-threo-DOPS, the industrial precursor of NE, was introduced as an agent to compensate the metabolic deficiency and thus improve the symptom. Although it is a gentle drug with mild effect, it reverses clinical symptoms relatively well without any side effects. Details of L-threo-DOPS treatment are described in another chapter by *Kondo*.

During clinical trials of L-threo-DOPS in chronic PD cases, changes and improvement in psychological sphere were noticed. Thus a new aspect of the disease for study today is the symptomatology of central NE deficiency in parkinsonism, which may include both motor and psychological symptoms such as depressive mood, emotional hypoactivity, and lowered level of attention.

§III. In parallel with going-on analysis of the underlying mechanism of each symptom, the interest of the author and of his colleagues has gradually shifted to the problem of the progressive nature of the disease. The difference in speed of progression or of worsening of the clinical picture in different age groups has been considered important. As an approach to this problem, a group of patients with disease onset at a young age was taken as a topic of interest.

In this chapter, juvenile parkinsonism (JP) is described by *Yokochi*, which includes both the cases in which the disease started below the age of twenty (cases of juvenile onset) and those in which it started between 21 and 40 (cases of younger onset or early starting cases). It was confirmed that these cases are of higher familial incidence and are much slower in progression when compared with PD. If the treatment, either pharmacological or surgical, is successfully applied, i.e., if it alleviates rigidity, secondary akinesia, tremor and primary akinesia, these younger starting cases can stay almost normal in motor and psychological function in working in jobs, social activities, and home life for ten to more than twenty years after the start of treatment. This may indicate that the basic pathological process is very slow in progression and also limited to the nigrostriatal DA system.

In parallel with analysis of the cases of juvenile or younger onset parkinsonism, the group of cases presenting dystonia and starting at the first decade of life is tentatively proposed to be a separate entity from JP, and will be described by *Segawa* under the title of hereditary progressive dystonia (HPD). These cases present diurnal fluctuation in severity of dystonia and markedly respond to the minimum dose of levodopa. Much interest is becoming focused on whether this is a separate condition from JP or represents a subtype of it.

In contrast to JP cases, generally speaking, PD cases starting after the age around sixty tend slowly but steadily to deteriorate within several to ten years, even after similarly successful initial results by medicine or surgery. This is presumed to be partly due to a faster degenerative process within the DA structures involved and also to widening or spreading of the pathological process to other related structures or functional systems such as NE- or 5HT-driven ones. Such difference between two groups, JP and PD, is interpreted as possibly related to the grade of aging of the brain in each patient, specially from studies on the pathology of the locus ceruleus and on the DßH activity of the brain. Differences in neuropathology of MPTP monkeys between the younger and older animals have been reported by Forno et al. (1988).

Progression in severity of each symptom can be interpreted as progressive degeneration of neurons involved in a certain neuronal structure or system. Worsening of rigidity, and Type-I and II akinesia in juvenile onset or early starting parkinsonism is attributable to progression of DA deficiency within the nigrostriatum, in which the pathology may start from the nigroputaminal pathway later spreading to the nigrocaudal one, as described elsewhere (Narabayashi, 1990b).

On the other hand, widening or spreading of morphological and chemical pathology to other additional structures may also contribute to the worsening of the whole clinical picture. In the classical PD cases, which start relatively late in life, pathology may start with nigrostriatal DA deficiency as in JP but later, in the longstanding progressive course, develop symptoms of NE deficiency with additional pathology in the locus ceruleus. Such understanding is gradually being supported by clinical observation, neurochemical and neuropathological analysis of brain tissue, and also by the results of clinical application of L-threo-DOPS, although it may take several more years yet for this understanding to be proved correct.

References

Forno LS, Langston JW, DeLanney LE, Irwin I (1988) An electron microscopic study of MPTP-induced inclusion bodies in an old monkey. Brain Res 448:150–157

Nakamura R, Narabayashi H (1976) Arrhythmokinesia in parkinsonism. In: Birkmayer W, Hornykiewicz O (eds) Advances in parkinsonism. Roche, Basle, pp 258–268

Narabayashi H (1990a) Surgical treatment in the levodopa era. In: Stern G (ed) Parkinson's disease. Chapman & Hall, London, pp 597–646

Narabayashi H (1990b) Clinical analysis of juvenile and classical parkinsonism and underlying pathophysiological mechanism. Presented at the XIth International Congress of Neuropathology, Kyoto, 1990

Correspondence: Prof. H. Narabayashi, Neurological Clinic, 5-12-8 Nakameguro, Meguro-ku, Tokyo 153, Japan.

Importance of juvenile parkinsonism in elucidating the pathogenesis of Parkinson's disease

M. Yokochi

Department of Neurology, Tokyo Metropolitan Institute for Neurosciences, Tokyo, Japan

Summary

In recent studies of Parkinson's disease, the importance of studying juvenile parkinsonism has been emphasized. The reason for this are as follows.

(1) Juvenile parkinsonism provides a crucial clinical model as a prototype of dopamine deficiency syndrome caused by nigro-striatal dysfunction. It might eliminate the factors related to aging from clinical and pathobiochemical findings.

(2) The case group of juvenile parkinsonism is situated between parkinson's disease with onset after middle age and dopa-responsive dystonia with onset in childhood. Without study of juvenile parkinsonism, further development of nosological discussion of the dopamine deficiency syndrome will not be possible.

(3) Interestingly, some cases have recently shown different pathological pictures in the substantia nigra from the traditional parkinsonian pathology of Lewy body pathology in autopsies. This evidence suggests that in juvenile parkinsonism, there is an underlying mechanism in the pathological process toward the lesion in the substantia nigra.

From these significant aspects, tracing of juvenile parkinsonism will contribute to finding the core of the pathophysiology and to determining the real pathogenesis of the dopamine deficiency syndrome.

In this paper, the clinical characteristics of epidemiology, symptomatology and prognosis, and pathological findings up to now, are presented. In addition, nosological discussion is developed on the basis of recent publications on juvenile parkinsonism.

Introduction

In previous reports (Yokochi, 1979a; Yokochi et al., 1984; Narabayashi et al., 1986) the author has detailed the characteristics of juvenile parkinsonism compared to old-age Parkinson's disease from the perspectives of epidemiology, symptoms, responses to levodopa, and prognosis of the illness. Recent autopsies have begun to shed light on the pathology. This paper discusses the disease and covers the newly clarified significance in the nosological aspect. (In this paper, the term *juvenile parkinsonism* will be used to describe Parkinson's disease with an early onset and such cases. Parkinson's diseases with onset after middle age and such cases will be designated as *Parkinson's disease*. The term *Parkinson's disease* will also be used for all cases with the early onset and the onset after middle age.)

The importance of tracing juvenile parkinsonism arises from tracing the pathogenesis of the dopamine deficiency syndrome which accompanies idiopathic nigral degeneration. Juvenile parkinsonism provides a crucial clinical model for the latter syndrome, because the former eliminates the factors related to aging from the clinical and patho-biochemical perspectives and also is believed to present a purer model of the dopamine deficiency syndrome. The other interest lies in whether different processes are conceivable for juvenile parkinsonism and Parkinson's disease upon establishment of the pathology. This interest merges with a further interest in nosological classification that ponders the differences and similarities with the pathogenesis of pediatric disorders such as Hereditary Progressive Dystonia with Diurnal Fluctuation (Segawa et al., 1976, 1988) or Dopa-Responsive Dystonia (Nygaard et al., 1988).

Barbeau (1986) has described Parkinson's disease as a symptom comlex of rigidity, hypokinesia, postural instability, and tremor, and as a clinical manifestation of an underlying dysfunction in the specific nigro-striatal system, and has noted the possible multifactorial origin of the disease within the scope of a specific nigro-striatal dopamine deficiency, but affected by external or internal environmental factors and not by a single factor. Interest exists in whether juvenile parkinsonism can become a specific example of this assumption.

Interest in juvenile parkinsonism along with hereditary progressive dystonia (HPD) arose early in Japan. At one time, the disease appeared to be particularly prevalent in Japan; however, as multiple cases have been reported recently in Europe and other countries, the disease is clearly not unique to Japan. In fact, these findings have created an international interest and generated a research topic for Parkinson's disease.

With the foregoing in mind, the author shall restate the clinical characteristics herein on the basis of the results of our own follow-up cases of juvenile parkinsonism collected in the author's original papers (described as *follow-up cases*) and introduce a summary of pathological observations from the autopsies. At the end of the report, development of a nosological discussion under current circumstances will be attempted.

Definitions

The author published a study on 40 cases of Parkinson's disease with onset before the age of 40 years (Yokochi, 1979a), and has tracked these cases for the last 15 years. The fundamental criteria for selecting these cases were that their major symptoms consisted of parkinsonism alone and that they had marked response to levodopa, aside from the particular points discussed later. These guidelines are vital when describing juvenile parkinsonism. The diagnostic definition at this time is simply "an illness with a younger onset showing major symptoms of parkinsonism and whose sufferers improve significantly upon levodopa treatment." The age classifications will be discussed further under the nosological discussion.

Prevalence

The percentage of onset below 40 years was 10.6% for 170 cases reported in the 1979 Yokochi report. For the combined survey covering several hospitals in Japan, the percentage was 11.5% among 424 cases. Hoehn and Yahr (1967) found 10.1% among 672 cases (under 40 years); Lima et al. (1987), 9.2% among 228 cases (under 40 years); Gershanik and Leist (1986), 3.8% among 336 cases (under 35 years); and Rajput et al. (1986), 4.8% among 505 cases (under 40 years). Some variance is evident, but many reports give percentages differing little from those obtained in Japan.

The male-to-female ratios for Yokochi (1979a), Gershanik and Leist (1986), Lima et al. (1987), and Ludin and Ludin (1989) are 25:15, 9:4, 9:12, and 11:12, respectively, showing equal representation, or a somewhat greater number of male cases.

Hereditary concentration

The average familial incidence for Parkinson's disease is said to be around 15%. Parkinson's disease is not considered to have high dependence of genetic factors, however, since the concordance between monozygotic twins appears to be low, and because the disease is considered to have a multifactorial origin with contributions from environmental factors. Nonetheless, families with juvenile onset cases experience a high rate of the illness. For instance, among the follow-up cases, 15 out of 35 families had familial incidence, with a high rate of approximately 43%. The younger a proband was, the higher the concentration in the family was found. The familial incidence rates related to onset age for the combined survey in Japan were 47.8% for 29 years and under, 29.6% for 30 to 39 years, 5.2% for 40 to 49 years, 6.7% for 50 to 59 years, 3.2% for 60 to 69 years, and 0% for 70 years and above. The comparison between ages under 40 years and those of 40 years or more was 36.7% vs. 4.8%. Recent reports on the incidence of familial cases of juvenile parkinsonism show rates of 46%, 23.1%, 9.5%, 33.3%, and 21.7%, respectively, for Barbeau and Pourcher (1982), Gershanik and Leist (1986), Lima et al. (1987), Quinn et al. (1987), and Ludin and Ludin (1989). Although contributions from environmental factors must be fully considered (Calne et al., 1987; Ludin and Ludin, 1989), genetic factors in juvenile parkinsonism may be quite important according to the data above. Among the 15 familial incidences, inheritances among the follow-up cases included six familial incidences among siblings, three familial incidences between siblings and other relatives, and six familial incidences between relatives. Families with patients of which one case was a parent numbered four out of 15. The foregoing strongly suggests autosomal recessive inheritance.

Symptoms

The following describes the clinical characteristics of juvenile Parkinsonism. As discussed at the beginning and for emphasis here, the fundamental symptoms do not deviate from symptoms of Parkinson's disease. From time to time, the following characteristics may be the only symptoms emphasized; however, if only such characteristics are marked, these indicate that another illness might be conceivable.

Tremor

Fewer than half the cases indicate regular, pill-rolling type tremor accompanying intrinsic plus hand at rest. Many cases indicate tremor of smaller amplitude and higher frequency only in posture and/or in action. Some cases demonstrate fine tremor, including the trunk, only when standing. Quinn et al. (1987) has estimated 46% as the characteristic tremor type, and 54% as the akinesia and rigidity type. In Parkinson's disease, symptoms frequently occur on one side of the body, but in juvenile parkinsonism, they characteristically shows symptoms on both sides throughout the illness, although some asymmetry may be recognized.

Rigidity and posture

Typical cogwheel rigidity is rare. The passive stretch response gives frequently tonic resistance accompanying irregular lapses. Symptoms of stooped and forward-bent postures which are believed to arise from a tonus imbalance in the trunk muscles, are mild or unaccompanied. Particularly under conditions of suitable treatment, many cases have no posture abnormality.

Akinesia and bradyphrenia

Without treatment for cases of moderate or more than moderate severity, the patient becomes bedridden, and at times, even raising the limbs becomes difficult. Even in such conditions, phonation and articulation remain relatively normal, and conversation is active. The masked face is also relatively unimpaired. As a point for careful observation, rather active movement is sometimes seen in the remaining motor functions. Ordinarily, levodopa treatment alleviates and improves bradykinesia and clumsiness. In the cases provided with optimum treatment, the severity of akinesia which regularly identified in Parkinson's disease, is usually scarcely in juvenile parkinsonism. Now, the symptomatological definition of the so-called akinesia has not been determined. We classify the symptom into four categories for symptom analysis: "hypokinesia", "loss of accessory movement", "initial hesitation", and "bradykinesia". The author believes that the inherent factors of akinesia are "hypokinesia", where although movement is possible if the patient has the will power, the movement becomes limited over time, and "loss of accessory movement", where although easily accomplished motor functions remain, rather purposeless casual body movements which a healthy subject would perform are not observed. The mild impression of akinesia in juvenile parkinsonism is derived from the mildness of these two disorders.

Moreover, the psychicakinesia or bradyphrenia resulting from the cognitive slowing accompanying impending concentration and apathy (Rogers et al., 1987) appears mild. Since juvenile parkinsonism shows marked improvement upon treatment with levodopa, the improvement of bradykinesia and clumsiness together with rigidity demonstrates that these symptoms are compensated by the supplementation of dopamine in the brain. Consequently, the pathogenesis of juvenile parkinsonism can be assumed to be chiefly dopamine deficiency, and particularly with a restriction to the nigro-striatal system. Parkinson's disease, however, is assumed to have more extensive pathological changes than those simply repaired by dopamine supplementation. The inherent pathology shows a large possibility of the added involvement of a mesocortico-limbic dopaminergic system disorder (Javoy-Agid et al., 1984) thought to be the cause of brady-phrenia and depression, and a noradrenaline system disorder, etc. Only three out of the twenty-two follow-up cases had WAIS scores of a total of 80 points or less. Lima et al. (1987) recognized the conditions of depression in seven cases out of 22, but recognized no mental dysfunctions including the stage IV cases of Hoehn and Yahr. Gershanik and Leist (1986) also found no mental dys-function or recognitive dysfunctions among the collected cases. These findings show either no or extremely light complications of mental disorders with juvenile parkinsonism. Some opinions (Ludin and Ludin, 1989; Lieberman et al., 1979) state, however, that mental disorders in Parkinson's disease drop in correlation with age only. In the event of studying only young patients, consideration should be given to this point.

Autonomic disturbances

The frequency and severity of complications such as seborrheic skin, constipation, hypotension, etc., are respectively low and mild. Lima et al. (1987) reported no complications in the collected cases, and Gershanik and Leist (1986) reported complications in only 15% of the cases.

Efficacy of levodopa and adverse effects

The dramatic effectiveness of levodopa was one of the first facts in outlining juvenile parkinsonism. A typical example would be that a patient off medica-tion for several days or more would be forced into a passive, bedridden position.

Administration of only 500 mg of levodopa allowed the patient to rise 15 to 30 minutes later, and to run in the hallway. Moreover, some cases have complications of various dopa-induced dyskinesia at peak doses. This type of effect almost certainly does not occur in elderly cases. Of course, not all cases are as dramatic as described above, but levodopa treatment must show a significant effect. The foregoing suggests, as noted earlier, that the subject cases consist of a dopamine-deficient illness, and that the pathological change is located in the nigro-striatal system. The phenomenon of this levodopa effect is also important from the diagnostic and differential diagnostic perspectives. Although levodopa treatment has a significant effect, complications arise as adverse effects sooner or later. These consist of up and down or wearing off of symptoms that accompany administration, and dopa-induced dyskinesia. The adverse effects are heavy, emerge relatively early during treatment compared to Parkinson's disease, and the frequency and severity of complications are severe. Quinn et al. (1987) noted that the complication rate of dyskinesia during a treatment period of six years reached 100%, while fluctuations reached 96% positive. Gershanik and Leist (1986) reported dyskinesia complications in 76% of the cases, which exceeded the 43% found for Parkinson's disease. In daily clinical practice, the patient and physician are absorbed in attempts to stabilize the fluctuations of the symptoms.

Another peculiarity compared to Parkinson's disease relates to the difference in the distribution and characteristics of dyskinesia, not just the frequency and severity (Yokochi and Nishimiya, 1985). Without exception, the oro-mandibula region is initially affected by the emergence of dyskinesia in Parkinson's disease. In juvenile parkinsonism, however, choreic movement or dystonia and sometimes ballism like motion occurs on the limbs, and involuntary movement in the oro-mandibula is rare. This evidence seems to reflect some differences in brain pathology, but the reasons remain unclear. According to the electromyogram analysis using the telemetry system, the conversion from parkinsonism to induced dyskinesia occurs over 1 to 2 minutes.

The author considered that characteristics of the levodopa effect might be shown pharmacokinetically in the blood, and subsequently measured dopa and its metabolites in the plasma over time after administration of levodopa (Yokochi, 1979b). The same amount of levodopa administration resulted in a concentration seven times higher in the plasma for juvenile parkinsonism compared to Parkinson's disease. A comparative study with control subjects was performed, and showed that the control group and juvenile Parkinsonism group were similar in plasma concentration, and only the Parkinson's disease group had a low concentration. This difference disappeared with the combination with decarboxylase inhibitor, however. This pharmacological finding may be part of the background related to the significant effect of levodopa for juvenile parkinsonism; the cause of the difference is not clear. Wakabayashi et al. (1988) showed that the Lewy body existed in the intestinal nerve plexus in Parkinson's disease, providing a suggestive commentary.

The autopsied Parkinson's disease brain shows a decrease in tyrosine hydroxylase activity and a decrease of biopterin, a coenzyme. Kondo et al.

(1986) reported that biopterin in the blood of cases with juvenile parkinsonism and of relatives assumed to be carriers maintained low values, and that a rise of biopterin by a tyrosine load was limited in juvenile parkinsonism and their relatives. This study also noted that the natural form of biopterin was effective toward various symptoms in these cases.

The combined administration of MAO-B inhibitor, deprenil and Ro19–6327 has shown an enhanced action for levodopa in the cases of juvenile parkinsonism. The investigation is currently underway for comparison with Parkinson's disease.

Prognosis of illness

Approximately 15 years have passed since the author began to track the follow-up cases. For the surviving cases, including three cases of over 40 years, the average duration of the illness is approximately 28 years. The average duration of the illness is 22 years for the 10 deceased cases. Lima et al. (1987), Ludin and Ludin (1989), and Gershanik and Leist (1986) reported the average number of years for collected cases as 19.8, 14.9, and 17.0, respectively. These figures are clearly longer than the average 7.4 ± 5.7 years for Parkinson's disease in Japan reported by Harada et al. (1983) and the 9 or 10 years reported by Diamond and Markham (1976). Moreover, prognostic investigations of our collected cases show that approximately 60% of surviving cases are either in the work force or homemakers. As long as appropriate treatment is provided, the prognoses are good (Yokochi, 1988a).

Pathology

Pathological findings related to juvenile parkinsonism are limited.

Ota and Miyoshi (1958) have reported three cases of juvenile onset displaying typical Parkinson symptoms, and in particular, the autopsy of a female case with an onset at 20 years and duration of 27 years. There is no description of the Lewy body, but the report stated no contradictions to the pathological impressions of Parkinson's disease. For a long time, this report was the sole autopsy report. Recently, Miyazawa et al. (1987) reported the autopsy of a sporadic case with an onset at 24 years. The distribution of lesions did not contradict that of Parkinson's disease, but the Lewy body was not found at any sections. This evidence pointed possibly to a different pathological mechanism than that for Parkinson's disease. Yokochi et al. (1984) reported autopsies of two cases with onset ages of 38 and six among the follow-up cases in the original report. The former case had no contradictions with Parkinson's disease; the latter also had no contradiction in the legions, but the substantia nigra showed a peculiar aspect. Mizutani et al. (1990) have recently reported the pathological findings in detail for the latter. Later we had the opportunity to perform five autopsies of three of the follow-up cases, one case of a sibling among the follow-up cases, and one case other than the follow-up cases, for a total of seven autopsies including the two above mentioned. The general findings have already been reported by Yokochi et al. (1988b) and Mizutani et al. (1988). The distribution of lesions for all of the cases had no contradictions with Parkinson's disease, but the conditions of the substantia nigra lesions were varied. We are currently studying the cases by classifying them into three categories: 1) typical Lewy body pathology cases with no contradiction with traditional Parkinson's disease (onset at *38 years*, 39 years, and 38 years); 2) Lewy-body-absent cases (onset at 14 years and 18 years); 3) peculiar substantia nigra pathology cases where in one case gliosis was quite mild and did not match the progressed years of the illness or the degree of the decrease of cell population (onset at 28 years) and in another case, the substantia nigra appeared hypoplasia or poor neuronal maturation with significantly deficient pigmentation and only mild gliosis (onset at *six years*; deceased at 39 years). The cases in italics were reported by the authors in 1984.

Gibb and Lees (1988) studied preserved brains and compared 12 cases with onset ages of 23 to 40 years (average 34 years) and 22 cases with onset ages of 70 years and above. In the substantia nigra, the decrease of the number of neurons and diminishment of melanin were the same for both groups. The juvenile group had a stronger decrease in cells containing melanin. This was assumed to be related to the length of illness, and the report determined that an inherent difference could not be found between the groups.

Cases described by Hunt (1917) consisting chiefly of pallidum lesions have long been considered juvenile parkinsonism. These cases can be easily and theoretically assumed to have completely different disorder distributions from Parkinson's disease and not to display levodopa effectiveness. For the foregoing reason, the cases of "progressive atrophy of the globus pallidus" identified by Hunt differ from the illness discussed in this paper. In fact, we have not experienced any case with a core of lesions other than the substantia nigra in the follow-up cases. This evidence is backed up by clinical findings that show the significant effect of levodopa for our follow-up cases.

Nosology

Development of the nosological discussion

Until recently, research on juvenile parkinsonism both in Japan and abroad has been limited to sporadic case reports and has not yet led to study of the concept of the illness. Scott and Brody (1971) reported 18 cases with an average onset age of 37.2 years for onesided progression over long periods averaging 22.2 years as a peculiar case group while tracking patients who mostly underwent a thalamotomy. History of high fever was described in more than half of the cases besides encephalitis.

"Juvenile parkinsonism" by the follow-up cases (1979) and "Hereditary Progressive Dystonia with Marked Diurnal Fluctuation" observed by Segawa et al. (1976) were reported independently. After these publications, active discussion began regarding the pathogenesis and nosology for both diseases in Japan. In response to discussion on the former, Barbeau and Pourcher (1981) reported the results of study in Canada at the 12th World Congress of Neurology (1981, Kyoto). It seems that this occasion marked the first step toward international interest.

Since the illness is called "juvenile" as a factor in the definition, establishing the range of the onset age becomes a problem. Yokochi established an onset age of under 40 years as a working hypothesis; therefore, there is no absolute meaning in the age itself. However, in later research reported by Barbeau and Pourcher (1982), Ishikawa and Miyatake (1985), Rajput et al. (1986), Quinn et al. (1987), Gibb and Lees (1987), Lima et al. (1987) and Ludin and Ludin (1989), the onset age was established as approximately 40 years for analysis. The author accepts one approach by Gershanik and Leist (1986), who adopted the part over 2SD of the average onset age from the onset age distribution of many cases of Parkinson's disease and determined the threshold at 35 years. Either way, it seems that the distribution of the onset age of Parkinson's disease is continuous and juvenile parkinsonism is in the lower end of the skewed deviation of the distribution of Parkinson's disease. Dividing cases into onset years of under 40 years is an approximation, since it is quite artificial to classify only by the onset age.

Subgroup establishment

Next, there is an approach for establishing subgroups for these case groups. Since the background to this approach probably accompanies each identification of pathogenesis, it may be important. Yokochi (1979a) presented the follow-up cases by dividing them into three groups in consideration of the reactivity to levodopa and the symptoms. The cases with complications such as factors of dystonia, inverted posture of the foot, and peculiar walks were designated as group III. This was a classification from the symptoms, but this group shared all cases of onset at extremely young ages under 16 years in the follow-up cases; namely, two at 6 years. Later, interest regarding the differences and similarities of this group and dopa-responsive dystonia arose. Ishikawa and Miyatake (1988) proposed classifying the groups as follows: I, the type that has the type of autosomal recessive inheritance and shows improvement of symptoms by sleeping; II, the type of juvenile onset of idiopathic Parkinson's disease; III, hereditary progressive dystonia that shows the dramatic diurnal fluctuations. In addition, Quinn et al. (1987) developed their own theory classifying onset cases of under-40-year-old patients as under or above 21 years of age, where the former is "juvenile parkinsonism," and the other is "young onset Parkinson's disease." Also, Barbeau and Pourcher (1982) and Lima et al. (1987) advocated classification by a "group mainly characterized by tremor" and a "group mainly characterized by rigidity and akinesia." Yamamura et al. (1968, 1973) and Sunohara et al. (1982, 1985) emphasized diurnal fluctuations in symptoms, reporting symptom groups having combined characteristics of both Segawa's reported cases and the juvenile parkinsonism reported by the author, and suggesting that they might be independent types.

Each type of classification has its own bases, however practically, it is definitely not simple to assign a classification for each individual case, and the classifications become artificial. Conditions change over the course of illness and with treatment. Consequently, as stated in the beginning, it is important for the time being to observe and analyze patiently from a variety of perspectives cases of onset at 40 years or under where levodopa shows significant effectiveness and which have no symptoms far contradictory to Parkinson's disease. It is essential to compile autopsy studies so that discussion develops from the basis of pathological knowledge.

Evidence for the independence of juvenile parkinsonism

Quinn et al. (1987) assessed four cases as juvenile parkinsonism with an onset at 21 years and under among 60 collected cases of an onset age of under 40 years. Because the cases were familial, and the affected relatives all experienced onset under 40 years, they hypothesized the cases as parkinsonism associated with heredity. The majority of the collected cases with an onset age between 21 years and 40 years described as young onset Parkinson's disease had long periods of illness. In addition, the clinical characteristics included the absence of dementia and marked response to levodopa treatment but with early emergence of dyskinesia and fluctuations. Nonetheless, the cases had a low frequency of familial onset and proposed the pathogenesis as degenerative Lewy body idiopathic Parkinson's disease, where the onset age was placed at the lower end of a skewed deviation of onset age distribution of Parkinson's disease while no additional factor causing early onset could be confirmed. Gershanik and Leist (1986) analyzed 13 cases of onset at 35 years and under obtained from 336 cases of Parkinson's disease, and stated an opinion similar to that of Quinn et al. (1987). Gibb and Lees (1988) recognized a small clinical difference between young- and old-onset cases of Parkinson's disease, but attributed this to physiological change with aging of the basal ganglia. The study also proclaimed both cases similar in Lewy body pathology under discussion of the pathology. Lima et al. (1987) and Ludin and Ludin (1989) are also negative on handling the disease independently. Rajput et al. (1986) hypothesized the association of environmental factors to onset, and discussed the potential exposure of the juvenile onset group to some factor in early childhood in the rural area. Barbeau and Pourcher (1981) stated that a peculiarity to the juvenile onset group consisted of a high rate of complications of metabolic disease (thyroid dysfunction, diabetes mellitus, hypertension). No reports supporting these abnormalities have appeared since then, however.

As described above, many opinions do not support the consideration of juvenile parkinsonism and Parkinson's disease separately. Certainly, the onset age distribution of Parkinson's disease does not have a double peak, and juvenile onset cases are relegated to the lower end of the skewed deviation of Parkinson's disease. The author has noted this early on. It seems inappropriate, however, to disregard clearly noteworthy clinical characteristics from this statistic alone, and to negate a hypothesized mechanism different in pathological process with the substantia nigra at the core. For instance, some cases may potentially create a spectrum joined to dopa-responsive dystonia, which is currently treated independently; i.e., not in conjunction for Parkinson's disease. From another perspective, in the proper clinical sense, the consideration that the clinical symptoms of old Parkinson's disease, juvenile parkinsonism, and dopa-responsive dystonia mutually suggest the same disease is strained. Dopa-responsive dystonia with the dystonia as its major characteristic and old Parkinson's disease can be thought of as different illnesses based on different processes. If so, juvenile parkinsonism must be significantly placed as a group between dopa-responsive dystonia and Parkinson's disease.

Dopa-responsive dystonia versus juvenile parkinsonism

The following continues discussion on the peculiarities and continuity of both illnesses. The differences lie in the symptoms and progress of treatment with levodopa. The cases of Segawa et al. (1976) show dystonia with lower limb including the trunk and neck, but no symptoms indicating parkinsonism. Particularly, bradykinesia or hypokinesia is quite mild or ordinarily absent and levodopa has a curable effect. Adverse effects of levodopa are nearly unrecognizable with daily treatment doses. In addition to these clinical findings, the prevalence has a gender difference; female cases are far more common, and familial cases often take an autosomal dominant trait. These differences suggest that the cases of dopa-responsive dystonia should be treated separately from juvenile parkinsonism. However, there are autosomal recessive family cases as well as autosomal dominant families (Ouvrier, 1978). Moreover, Nygaard and Duvoisin (1986) reported three autosomal dominant familial cases showing complications of parkinsonism with bradykinesia and tremor. Consequently, Nygaard and Duvoisin called the illness "hereditary dystonia-parkinsonism syndrome of juvenile onset". The inclusion of these cases cannot clinically negate the continuity with juvenile parkinsonism. Since no autopsies have been performed for the dystonia cases, the following is speculative: the effectiveness of dopamine supplementation therapy suggests a dopamine deficiency disorder of the nigro-striatal system. As such, these illnesses can be considered to exist on the continuous spectrum from dopa-responsive dystonia through juvenile parkinsonism to Parkinson's disease.

Conclusions

The following opinion regarding nosological theory on juvenile parkinsonism concludes this paper.

The focus on juvenile parkinsonism among Parkinson's disease has followed a large effort in symptomatology to develop levodopa treatment. At the time when the results of levodopa treatment could not be considered, no doubt many cases were sealed without the benefit of diagnosis. Today, however, absolute evidence for classification of juvenile parkinsonism as a disease entity other than Parkinson's disease does not exist. The quality and magnitude of the symptoms and treatment effectiveness show a continuum with Parkinson's disease, and a rigid difference between familial cases and sporadic cases cannot be identified. The pathological findings obtained thus show the chief site of pathological change in a restricted region of the brain. In other words, the illness is definitely a deficiency of dopamine, a vital transmitter of the basal ganglia, caused by the dysfunction of the substantia nigra. The results in no way contradict clinical evidence of functional repair through dopamine supplementation treatment. From the perspective of an idiopathic dopamine deficiency syndrome, the foregoing promotes a pathological mechanism common to Parkinson's disease, juvenile parkinsonism, and dopa-responsive dystonia. A further clinical review, however, recalls rigid differences in symptoms and in the influence of hereditary elements; hence, there is difficulty in considering the spectrum as one syndrome.

Consideration of differences and similarities with the existing illness outline and of the independent nature of the illness forces a cyclical argument. Nonetheless, juvenile parkinsonism remains a prototypical disease in the nigrostriatal dysfunction, and provides the best clinical model for research of the pathogenesis of the dopamine deficiency syndrome.

References

Barbeau A, Pourcher E (1981) Studies on the etiology of Parkinson's disease. In: Abstracts, 12th World Congress of Neurology. International Congress Series: p 548. Excerpta Medica, p 132

Barbeau A, Pourcher E (1982) New data on genetics of Parkinson's disease. Can J Neurol Sci 9:53–60

Barbeau A (1986) Parkinson's disease: clinical features and etiopathology. In: Vinken PJ, Bruyn GW, Klawans HL (eds) Handbook of clinical neurology, vol 5(49). Elsevier Science Publisher, pp 87–152

Calne S, Schoenberg B, Martin W, Uitti RJ, Spencer P, Calne DB (1987) Familial Parkinson's disease: possible role of environmental factors. Can J Neurol Sci 14:303–305

Diamond SG, Markham Ch H (1976) Present mortality in Parkinson's disease: the ratio of observed to expected deaths with a method to calculate expected deaths. J Neural Transm 38:259–269

Gershanik OS, Leist A (1986) Juvenile onset Parkinson's disease. Adv Neurol 45:213–216

Gibb WR, Lees AJ (1987) The progress of idiopathic Parkinson's disease is not explained by age-related changes. Clinical and pathological comparison with post-encephalitic parkinsonian syndrome. Acta Neuropathol (Berl) 73:195–201

Gibb WRG, Lees AJ (1988) A comparison of clinical and pathological features of young- and old-onset Parkinson's disease. Neurology 38:1402–1406

Harada H, Nishikawa S, Takahashi K (1983) Epidemiology of Parkinson's disease in a Japanese City. Arch Neurol 40:151–154

Hoehn MM, Yahr MD (1967) Parkinsonism: onset, progression, and mortality. Neurology 17:427–442

Hunt JR (1917) Progressive atrophy of the globus pallidus. Brain 40:58–148

Ishikawa A, Miyatake T (1985) Pathophysiology of juvenile parkinsonism-regarding juvenile parkinsonism with autosomal recessive trait. Jpn J Neuropsychopharmacol (Jpn) 7:809–815

Ishikawa A, Miyatake T (1988) Juvenile parkinsonism. In: Miyatake T (ed) Niigata Symposium Series on Neurology, No5. Parkinsonism (Jpn). Kagaku Hyoronsha, Tokyo, pp 22–49

Javoy-Agid F, Ruberg M, Taquet H, Bokobza B, Agid Y, Gaspar P, Berjer P, N'Guyen-Legros J, Albarez C, Gray F, Esconelle R, Scatton B, Roquier L (1984) Biochemical neuropathology of Parkinson's disease. Adv Neurol 40:189–198

Kondo T, Narabayashi H, Nagatsu T, Yamaguchi T, Sawada M (1986) Effects of tyrosine administration on plasma biopterin in patients with juvenile parkinsonism and their relatives. Adv Neurol 45:217–222

Lieberman A, Dziatolowski M, Kupersmith M, Serby M, Goodgold A, Korein J, Goldstein M (1979) Dementia in Parkinson's disease. Ann Neurol 6:355–359

Lima B, Neves G, Nora M (1987) Juvenile parkinsonism: clinical and metabolic characteristics. J Neurol Neurosurg Psychiatry 50:345–348

Ludin SM, Ludin HP (1989) Is Parkinson's disease of early onset a separate disease entity? J Neurol 236:203–207

Miyazawa Y, Abe N, Ohtoh T (1987) An autopsy case of juvenile parkinsonism. Neurol Med (Jpn) 26:578–583

Mizutani Y, Sugita Y, Kosaka K, Yokochi M, Satoh T, Narabayashi H (1988) Juvenile parkinsonism: neuropathological findings in 7 cases. In: Abstract for 29th General Meeting of Societa Neurologica Japonica (Jpn), p 271

Mizutani Y, Yokochi M, Oyanagi S (1991) Juvenile parkinsonism: a case with first clinical manifestation at the age of six years and with neuropathological findings suggesting a new pathogenesis. Clin Neuropathol (Berl) (in press)

Narabayashi H, Yokochi M, Iizuka R, Nagatsu T (1986) Juvenile parkinsonism. In: Vinken PJ, Bruyn GW, Klawans HL (eds) Handbook of clinical neurology 5 (49). Elsevier Science Publishers, pp 153–165

Nygaard TG, Duvoisin RC (1986) Hereditary dystonia-parkinsonism syndrome of juvenile onset. Neurology 36:1424–1428

Nygaard TG, Marsden CD, Duvoisin RC (1988) Dopa-responsive dystonia. Adv Neurol 50:377–384

Ota U, Miyoshi S (1958) Familial paralysis agitans juveniles: a clinical, anatomical and genetic study. Folia Psychiat Neurol Jpn 12:112–121

Ouvrier RA (1978) Progressive dystonia with marked diurnal fluctuation. Ann Neurol 4:412–417

Quinn N, Critchley P, Marsden CD (1987) Young onset Parkinson's disease. Movement Disorders 2:73–91

Rajput AH, Uitti RJ, Stern W, Laverty W (1986) Early onset Parkinson's disease and childhood environment. Adv Neurol 45:295–297

Rogers D, Lees AJ, Smith E, Trimble M, Stern CM (1987) Bradyphrenia in Parkinson's disease and psychomotor retardation in depressive illness. An experimental study. Brain 110:761–776

Scott RM, Brody JA (1971) Benign early onset Parkinson's disease: a syndrome distinct from classic postencephalitic parkinsonism. Neurology 21:366–368

Segawa M, Hosaka A, Miyagawa F, Nomura Y, Imai H (1976) Hereditary progressive dystonia with marked diurnal fluctuation. Adv Neurol 14:215–223

Segawa M, Nomura Y, Tanaka S, Hakamada S, Nagata E, Soda M, Kase M (1988) Hereditary progressive dystonia with marked diurnal fluctuation – consideration on its pathophysiology based on the characteristics of clinical and polysomnographical findings. Adv Neurol 50:367–367

Sunohara N, Mano Y, Toyoshima E, Ando K, Satoyoshi E (1982) Juvenile parkinsonism with marked diurnal fluctuation-about new findings. Clin Neurol (Jpn) 22:101–105

Sunohara N, Mano Y, Ando K, Satoyoshi E (1985) Idiopathic dystonia – parkinsonism with marked diurnal fluctuation of symptoms. Ann Neurol 17:39–45

Wakabayashi K, Takahashi H, Takeda S, Ohama E, Ikuta F (1988) Parkinson's disease: the presence of Lewy bodies in Auerbach's and Meissner's plexuses. Acta Neuropathol 76:217–221

Yamamura Y, Iida M, Ando K, Sobue I (1968) A juvenile familial disorder with rigido spasticity, bradykinesia and minor dystonia alleviated after sleep. Clin Neurol (Jpn) 8:233–243

Yamamura U, Sobue I, Ando K, Iida M, Yanagi T, Kono C (1973) Paralysis agitans of early onset with marked diurnal fluctuation of symptoms. Neurology 23:239–244

Yokochi M (1979a) Juvenile Parkinson's disease, part I. Clinical aspects. Adv Neurol Sci (Jpn) 23:1048–1059

Yokochi M (1979b) Juvenile Parkinson's disease, part II. Pharmacokinetic study. Adv Neurol Sci (Jpn) 23:1060–1073

Yokochi M, Narabayashi H, Iizuka R, Nagatsu T (1984) Juvenile parkinsonism-some clinical, pharmacological and neuropathological aspects. Adv Neurol 40:407–413

Yokochi M, Nishimiya J (1985) DOPA induced dyskinesia. Adv Neurol Sci (Jpn) 29:276–286

Yokochi M (1988a) Long-term follow-up study in patients with juvenile parkinsonism. In: Nakanishi T (ed) Annual report of the Research Committee of CNS degenerative disease. The Ministry of Health and Welfare of Japan, pp 268–272

Yokochi M, Mizutani Y, Narabayashi H, Tsuboi H (1988b) Long-term follow-up study in patient with juvenile parkinsonism. In: Book of Abstracts for 9th International Symposium on Parkinson's disease (Jerusalem), p 70

Correspondence: Dr. M. Yokochi, Department of Neurology, Tokyo Metropolitan Institute for Neurosciences, 2-6 Musashidai, Fuchu-si, Tokyo-to 183, Japan.

Dopa-unresponsive pure akinesia or freezing

H. Imai

Department of Neurology, Juntendo University School of Medicine, Tokyo, Japan

Summary

Barbeau (1972) first described "pure" akinesia without rigidity and tremor responsive to L-dopa therapy (akinesia due to striatal dopamine deficiency). Since 1974, Narabayashi and the present author described cases with pure akinesia unresponsive to L-dopa treatment as a new condition. This condition exhibits only the freezing symptom, which is a breakdown of repetitive voluntary movements emerging through festination or suddenly, e.g., freezing of gait, micrographia and inaudible speech. Kinésie paradoxale is always accompanied by this type of akinesia. The author suggested that the main pathological structure of the condition is different from the nigrostriatal dopaminergic system and that the condition is different from Parkinson's disease. L-threo-DOPS, a synthetic norepinephrine (NE) precursor, had a mild-to-moderate effect on some cases with freezing, and the NE hypothesis for freezing was proposed. Up to the present, more than 20 cases with this condition have been known in our clinic. All cases were sporadic and slowly progressive, and some had been followed for as long as ten years, still without rigidity and tremor. Slight muscular hypotonia was observed in the extremities. Slowly progressive ophthalmoplegia appeared later in several cases in which progressive supranuclear palsy (PSP) was strongly suggested despite no axial dystonia. Autopsy cases associated with this condition have been reported in Japan and pathologically revealed PSP or pallido-nigro-luysial atrophy. The nosological position and responsible lesion sites of this condition are discussed.

Definition of akinesia

Akinesia, the core symptom of parkinsonism, is a comprehensive term used in the daily practice of neurology. The definition of akinesia proposed by an ad hoc Committee in the Research Group on Extrapyramidal Disorders of the World Federation of Neurology (Lakke, 1981) is as follows:

"Akinesia" is a disorder characterized by poverty and slowness of initiation and execution of willed and associated movements and difficulty in changing one motor pattern to another, in the absence of paralysis. This may include an inability to sustain repetitive movements and difficulty in performing simultaneous motor acts and may vary in severity

from slight (sometimes called hypokinesia) to severe and complete immobility. The term bradykinesia should be reserved for slowness in the execution of movements.

More concretely, akinesia primarily includes: 1) mask-like face, 2) loss of finger dexterity, 3) loss of restless spontaneous movements of normals while sitting (Schwab and England, 1968), 4) loss of arm swing in walking, 5) petit pas gait, 6) micrographia, 7) loss of speech volume, tendency to speed up (festinating speech), and 8) freezing of gait (Schwab and Zieper, 1965). James Parkinson (1817) in his original essay pointed out the festinating gait (scelotyrbe festinans) as one of the two pathognomonic symptoms in patients with shaking palsy.

Though all of these symptoms are characteristic in parkinsonism, they have been considered heterogeneous in the pathophysiologic mechanism. How different is poverty of movement from slowness of movement? What is the difference in the mechanism between initiation and execution of movement? Festination in repetitive movements seems to have a different mechanism from other brady- and hypokinesia. "What is the nature of akinesia in parkinsonism?" is still a most challenging theme of research in kinesiology and clinical neurology.

Freezing and kinésie paradoxale

In order to further refine the meanings of akinesia, it would be appropriate to describe here in detail the freezing symptom and "kinésie paradoxale" (Souques, 1921) or "kinesia paradoxa" (Bing, 1923), which accompanies freezing. The freezing symptom is a breakdown of repetitive voluntary movements emerging through festination or suddenly, e.g., freezing of gait, micrographia and inaudible speech. As the representative of freezing, freezing of gait occurs either at the start (start hesitation), through festination of gait or suddenly when turning or going through the narrow corridor or door, or by emotional excitement. The patient cannot advance as if the soles of his or her feet are glued to the floor. He or she is embarrassed and frequently falls forward like a log. It is clearly different from paretic or ataxic gait, and is inherently is not a constant but rather a transient symptom.

Kinésie paradoxale was termed by Souques (1921). It is defined as a sudden total loss or overcoming of akinesia under special circumstances or stimulus conditions. For example, a patient is able to overcome the freezing of gait by stepping over an obstacle. He can walk fairly normally over adhesive tapes placed 18 to 36 inches apart on the floor and go smoothly up and down the staircase. These are appropriate visual (rhythmic) stimuli which induce kinésie paradoxale. Auditory stimuli, e.g., metronomic sounds and words of command, are also often effective. Freezing and kinésie paradoxale are phenomena accompanying each other and are easily reproducible, though both are transient and "episodic". Schwab and Zieper (1965) described episodes of complete loss of akinesia of severely disabling parkinsonian patients under emergency, extraordinary stimulation such as a flood or fire, but it lasted only a minute or two.

Tremor, rigidity versus akinesia

It has been widely accepted that cardinal symptoms of Parkinson's disease are related to the motor system. Tremor, rigidity and akinesia are called the motor triad, and the term "postural disorders" is frequently added to the triad (the motor tetrad). Both tremor and rigidity are, in the usual sense, simple and apparent signs easily objectified by surface EMG recording. Akinesia and postural disorders are signs in a dimension different from rigidity and tremor. Thus, classification by Martin (1967) is prominent for systematically under-standing the motor triad or tetrad in parkinsonism (Table 1). He applied the idea of "positive" and "negative" symptoms in neuropsychiatry, described by Hughlings Jackson (1884), to basal ganglia disorders. Martin, of course, layed emphasis on the negative symptoms, consisting of postural disorders and akinesia. Positive symptoms, i.e., rigidity and tremor, were considered to be important as presumed causes of negative symptoms. Existence of secondary akinesia due to rigidity and tremor (2a of Table 1) was confirmed by stereotaxic thalamotomy. Martin insisted that postural disorders (disorders of postural fixation and equilibrium) but not akinesia were the core of negative symptoms after analysis of postencephalitic parkinsonism. However, the relationship between akinesia and postural disorders remains to be clarified. Gait is a willed movement, but is continuously supported by automatic postural reactions and adjustments.

Dopa-responsive pure akinesia

A number of neurologists have experience that there are cases with parkinsonism, in which akinesia is not so improved through reducing the grade of muscular hypertone by stereotaxic thalamotomy. There are also many cases in whom the amount of akinesia is in no way connected with the amount of rigidity and tremor. Barbeau (1972) described an *almost* pure akinetic syndrome without rigidity and tremor as a subgroup of Parkinson's disease, which he found in 15% of his clinic parkinsonian population. On the basis of clinico-physiological analysis, decrease in urinary dopamine excretion and a good response to L-dopa therapy in such patients, it was concluded that the underlying mechanism of this akinesia is essentially the same as that of Parkinson's disease, i.e., striatal dopamine deficiency. He suggested that defect

Table 1. Motor symptoms in parkinsonism

1 Positive symptoms (release phenomena) rigidity, tremor
2 Negative or deficiency symptoms
a secondary to positive symptoms
b primary negative symptoms

(according to Martin, 1967)

in motor initiative – slow initiation time, decreased kinetic motivation and diminished associated movements – is the most fundamental component of akinesia, corresponding to presumably the dopaminergic set and trigger functions of the striatum.

Dopa-induced akinesia or freezing

Barbeau (1972, 1976) and Ambani and Van Woert (1973) pointed out akinesia paradoxica (hypotonic freezing) and start hesitation of gait respectively, as a serious side effect of long-term L-dopa therapy. This can be called dopa-induced akinesia or freezing. Its cumulative prevalence increased to involve 55% of the patients after six years of therapy, and data from episodes of hypotonic freezing have indicated that it is not accompanied by a drop in plasma dopa levels (Barbeau, 1976). Thus, Barbeau considered hypotonic freezing to be a norepinephrine (NE)-related phenomenon, and its increased incidence with prolonged L-dopa therapy as an indicator of the involvement of areas of the brain innervated from the locus ceruleus.

Recently, Quinn et al. (1989) reported an interesting case of dopa-responsive pure akinesia due to Lewy body Parkinson's disease with pathology. Although the initial symptoms and state of the case were not clearly described, "with L-dopa therapy the patient immediately developed drug-induced dyskinesia. Wearing-off and freezing became noticeable after 4 years of treatment, and on-off fluctuations persisted". Thus, in this case dopa-induced akinesia or freezing might be combined with dopa-responsive akinesia.

Dopa-unresponsive pure akinesia or freezing

1. Clinical features

Since 1974, Narabayashi and the author (1974, 1976, 1980, 1986) described cases with pure akinesia or freezing without rigidity and tremor and unresponsive to L-dopa treatment as a new condition or specific form of parkinsonism. This condition exhibited only the above-described freezing symptom, which was a breakdown of repetitive voluntary movements that occurred through festination or suddenly, e.g., freezing of gait, micrographia, and festinating speech. Kinésie paradoxale always accompanied this type of akinesia. This condition was considered quite different from the above-described cases of dopa-responsive pure akinesia, in whom defect in motor initiative or slow initiation time appeared to be the most prominent feature (Barbeau, 1972).

Usually freezing was most markedly recognized in walking, and less but still obviously observed sooner or later in repetitive arm movements such as teeth-brushing and writing and also in speech. Finger tapping response to

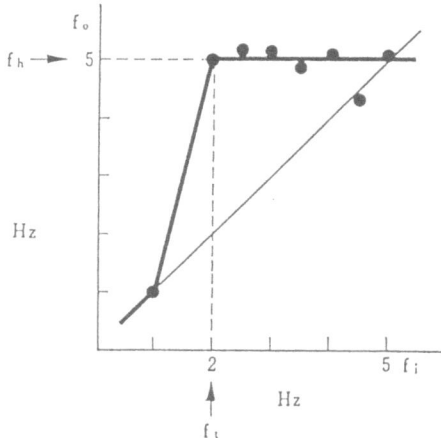

Fig. 1. Tapping response of a patient with dopa-unresponsive pure akinesia. The typical hastening phenomenon occurs when the signal rate is over 2 Hz. fi input (signal) frequency; fo output (response) frequency; ft transitional (input) frequency; fh hastening (output) frequency

periodic sound signals was examined in these patients. When the signal rate for tapping was over 2 or 2.5 Hz, they started to show hastening of repetition, converging into about 5 Hz (Fig. 1), i.e., the "hastening phenomenon" reported by Nakamura et al. (1976). This pattern was quite similar to that of patients with Parkinson's disease. Retropulsion was observed in all cases but the disturbance of other postural reactions was indefinite, varying case by case. Neither weakness nor ataxia was observed.

L-dopa treatment, either L-dopa alone up to 3 g or 600 mg of L-dopa plus 60 mg of carbidopa daily, was not at all effective against the freezing symptom and sometimes made it rather worse. HVA and MHPG levels in the cerebro-spinal fluid (CSF) of the four examined patients with this type of pure akinesia were not below normal. Thus, the author strongly suggested that the main pathological structure of the condition is different from the nigrostriatal dopaminergic system and the condition is different from Parkinson's disease (Imai, 1980).

Freezing symptoms slowly and progressively worsened. To date, more than 20 cases have been known in our department, and some had been followed up for as long as ten years, elevated to Yahr's stage 4 or 5, still without rigidity. All cases were sporadic and the age at disease onset was usually around fifty to seventy years, with only one exception of 32 years.

Accompanying neurological symptoms and signs of cases with this condi-tion were as follows. Slight muscular hypotonia in the extremities was observed in about two-thirds of the cases with no signs of cerebellar ataxia. Vertical gaze and convergence palsy, horizontal gaze nystagmus and apraxia of eyelid opening were noted in approximately one-third of the cases. Slight emotional incontinence and hyperreflexia were also noticed in approximately one-fourth of them.

A detailed neurootological examination, which was performed in recent cases (Imai et al., 1986), revealed convergence nystagmus, square wave jerks and flutter-like oscillations in some cases. Optokinetic nystagmus (OKN) was diminished in all examined cases, and vertical OKN, especially downward OKN, was markedly decreased, although these patients had no apparent downward gaze palsy. All patients showed "saccadic" pursuit eye movements during the eye tracking test. Slowly progressive supranuclear ophthalmoplegia with the medial longitudinal fasciculus syndrome appeared later in several patients in whom progressive supranuclear palsy (PSP) was strongly suggested although they had no axial dystonia.

A detailed neuroradiological study was carried out by pneumoencephalography (PEG) including tomography in the seventies (Imai, 1980), especially concerning fine cerebellar and brainstem structures in order to ascertain the pathology of the condition. Interesting localized atrophies were revealed, including asymmetry and/or enlargement of posterior superior recesses of the IVth ventricle and amplification of lateral walls of the IVth ventricle to the aqueduct cerebri. Recently, CT scan showed in most cases mild-to-moderate cerebellar cortical atrophy and dilatation of the IIIrd and lateral ventricles.

After our first paper in 1974, numerous clinical cases with this type of pure akinesia have been reported from all over Japan (e.g., Kurihara et al., 1977; Tokiguchi et al., 1978; Nakagawa et al., 1987) and recently from abroad (Fénelon et al., 1989).

2. L-threo-DOPS treatment against freezing

Narabayashi et al. (1981) introduced L-threo-3,4-dihydrophenylserine (L-threo-DOPS), a synthetic NE precursor, as an agent specifically effective on the freezing symptom. The idea was based on the finding that the activity of dopamine-β-hydroxylase (DBH), NE synthesizing enzyme, in the CSF was lowered more markedly in the later stage of Parkinson's disease than in the earlier stage. Therefore, the appearance and progression of the dopa-unresponsive freezing symptom in the later stage was assumed to be due to the deficiency of NE (NE hypothesis for freezing). In Japan, L-threo-DOPS is now commercially available against freezing in parkinsonism, and it has had mild-to-moderate effect on some cases with pure akinesia (Kondo, 1984; Yamamoto et al., 1985; Narabayashi et al., 1986). Kondo will discuss on this in detail in another chapter of this monograph. However, its long-term efficacy and pharmacological mechanism of action seem to remain unclear.

Other drug trials conducted to ameliorate this condition have been reported to be effective in a few cases: amantadine (Kurihara et al., 1977), L-5-HTP, the precursor of serotonin (Tokiguchi et al., 1978) and maprotiline, a tetracyclic antidepressant (Iwasaki et al., 1989). However, it seems that these effects might be only temporary or equivocal so far.

3. Autopsy reports associated with this condition

No case of our series has yet become available for autopsy in our department. Takahashi et al. (1977) reported two autopsy cases, showing severe akinesia, no rigidity in the limbs, no tremor but retropulsion, upward gaze palsy, dysarthria, dysphagia and later, nuchal stiffness. Both cases pathologically revealed pallido-nigro-luysial atrophy (PNLA, reported by Contamin et al., 1971) associated with degeneration of the centrum medianum of the thalamus.

Homma et al. (1987) reported an autopsy case at age 65, who had had a totally 11 years duration of illness and was initially diagnosed as having pure akinesia without rigidity and tremor and with no effect from L-dopa therapy. Eight years after the onset of freezing of gait, vertical oculomotor palsy, pseudobulbar palsy and dementia were added without nuchal dystonia. She was then diagnosed as having PSP, which was confirmed pathologically. The same group (Yuasa et al., 1987) reported five cases, including Homma's autopsied one, who were all initially diagnosed as having dopa-unresponsive pure akinesia, but during more than ten years were changed to be clinically diagnosed as having PSP.

PSP and PNLA, though the latter is a very rare disorder, clinically resemble each other closely. Differential diagnosis between the two definitely requires pathological examination.

Discussion

The nosological position of this condition or syndrome, probably as a degenerative disease or diseases, still remains uncertain. However, long-term observations for more than ten years of this condition and its autopsy studies have led us to the idea that PSP and its asociated conditions, i.e., atypical PSP (Davis et al., 1985), forme fruste of PSP (Yuasa et al., 1987) and PNLA, may be part of underlying disorders of dopa-unresponsive pure akinesia (Imai, 1980; Imai et al., 1986; Homma et al., 1987; Yuasa et al., 1987). In other words, this type of pure akinesia can be an initial symptom-complex of the "PSP group" in the clinical course. Findings of detailed neurootological examination and PEG study can retrospectively be regarded as supportive evidence for diagnosing the PSP group. However, it is too difficult to speculate the responsible lesion site of "freezing" from PSP pathology, which shows a "multiple" system atrophy. A reasonable speculation from the pathology of PSP back to the clinic of this type of pure akinesia may be on "without rigidity and tremor", because rigidity and tremor due to loss of the nigrostriatal dopaminergic neurons in PSP might be masked again by the pallido-luysial degeneration in PSP similar to the state of patients with Parkinson's disease after stereotaxic pallidotomy. This speculation can also explain the "unresponsiveness to L-dopa".

The idea that this type of pure akinesia is an *initial* symptom of PSP in a broad sense seems inharmonious with the NE hypothesis for freezing in

Parkinson's disease (Barbeau, 1976; Narabayashi et al., 1981). This is because: 1) increase in freezing symptom and involvement of the NE system appear in the *later* stage of Parkinson's disease, and 2) contrary to severe involvement of the dopamine neurons in the substantia nigra, involvement of the NE neurons in the locus ceruleus was reported to be minimal in PSP (Homma et al., 1987) and PNLA (Takahashi et al., 1977). However, any conclusion should be carefully postponed until after future study.

Meanwhile, this condition can be regarded as either a symptom-complex or a syndrome having a stereotypic clinical course. The condition clinically consists of three components: 1) freezing, 2) without rigidity and tremor, and 3) unresponsiveness to L-dopa. The latter two are modifying components to freezing, the first component. If we then focus on only freezing and neglect the accompanying symptoms or the clinical course, several other conditions can be pointed out as underlying disorders of this condition, e.g., multiple cerebral infarction involving the frontal lobes and/or basal ganglia, Binswanger's disease and normal pressure hydrocephalus. These conditions mainly produce gait disorder. Recently, "lower body parkinsonism (LBP)" of vascular etiology has been proposed, the main symptom of which is freezing of gait with no or only minimal upper limb involvement and with poor response to L-dopa (FitzGerald and Jankovic, 1989). Symptomatologically, considerable overlap exists between the following various gait disorders: parkinsonian gait, freezing of gait, démarche à petit pas and apraxia of gait. FitzGerald and Jankovic have suggested that disconnection between basal ganglia and the supplementary motor area is responsible for LBP. Nagasaki et al. (1981) presented their view that the hastening phenomenon (Nakamura et al., 1976) is correlated with striatal lesions on brain CT scan of patients with cerebrovascular disease. Other conditions, which have been reported to show freezing of gait and kinésie paradoxale are carbon monoxide (CO) intoxication (Imamura et al., 1979; Senoh et al., 1982) and Wilson's disease (Nakane et al., 1986). Cases with CO intoxication were found to have bilateral lucencies of the globus pallidus on CT scan consistent with necrotic lesions. These patients showed little or no clinical response to levodopa. CT scan of a case with Wilson's disease with kinésie paradoxale disclosed low densities in the bilateral lenticular nuclei. Therefore, it is suggested that the lenticular nucleus, pallidum and/or striatum, may be lesion sites responsible for freezing. It is also compatible with the discussion above on the pathology of PSP and PNLA, i.e., the pallidal degeneration.

Science repeatedly progresses by setting up hypotheses and then verifying them. The truth underlying the freezing symptom still lies deep inside the brain, like the state of the elephant approached for the first time by the six blind men of Indostan. For further progress, combined multi-disciplinary approaches are essential, including neuropsychology, neuroimaging and chemical neuropathology.

References

Ambani LM, Van Woert MH (1973) Start hesitation – a side effect of long term levodopa therapy. N Engl J Med 288:1113–1115

Barbeau A (1972) Contribution of levodopa therapy to the neuropharmacology of akinesia. In: Siegfried J (ed) Parkinson's disease. Hans Huber, Bern, pp 151–174

Barbeau A (1976) Six years of high-level levodopa therapy in severely akinetic parkinsonian patients. Arch Neurol 33:333–338

Bing R (1923) Über einige bemerkenswerte Begleiterscheinungen der "extra-pyramidalen Rigidität" (Akathisie-Mikrographie-Kinesia paradoxa). Schweiz Med Wschr 53: 167–171

Contamin F, Escourolle R, Nick J, Mignot B (1971) Atrophie pallido-nigro-luysienne. Syndrome akinetique avec palilalie, rigidité oppositionnelle et catatonie. Rev Neurol 124:107–121

Davis PH, Bergeron C, McLachlan DR (1985) Atypical presentation of progressive supranuclear palsy. Ann Neurol 17:337–343

Fénelon G, Ziegler M, Mahieux F, Guillard A (1989) Freezing of gait and resistance to dopaminergic therapy. In: Abstracts of Alzheimer's and Parkinson's Diseases. Kyoto, Nov 6–10, p 110

FitzGerald PM, Jankovic J (1989) Lower body parkinsonism: evidence for vascular etiology. Movement Disorders 4:249–260

Homma Y, Takahashi H, Takeda S, Ikuta F (1987) An autopsy case of progressive supranuclear palsy showing "pure akinesia without rigidity and tremor and with no effect by L-dopa therapy (Imai)". Brain Nerve (Tokyo) 39:183–187

Imai H (1980) Syndrome of pure akinesia or freezing phenomenon without rigidity and tremor and with no effect by L-DOPA therapy. Adv Neurol Sci (Tokyo) 24:838–848

Imai H, Narabayashi H (1974) Akinesia. Adv Neurol Sci (Tokyo) 18:787–794

Imai H, Narabayashi H, Sakata E (1986) "Pure akinesia" and the later added supranuclear ophthalmoplegia. Adv Neurol 45:207–212

Imamura S, Okajima T, Izumi J, Izuno R (1979) Kinésie paradoxale occurred after carbon monoxide intoxication. Clin Neurol (Tokyo) 19:429–433

Iwasaki S, Narabayashi Y, Takakusagi M, Iwasaki A, Hamaguchi K (1989) Maprotiline therapy for freezing of Parkinson's disease. In: Abstracts of Alzheimer's and Parkinson's Disease. Kyoto, Nov 6–10, p144

Jackson JH (1884) Evolution and dissolution of the nervous system (Croonian lectures). In: Taylor J (ed) Selected writings of John Hughlings Jackson, vol 2. Staples Press, London, 1985, pp 45–75

Kondo T (1984) D,L-threo-3,4-dihydroxyphenylserine (D,L-threo-DOPS) treatment on the patients with Parkinson's disease or pure akinesia. Clin Neurol (Tokyo) 24:280–288

Kurihara T, Terao A, Araki S (1977) Effective amantadine therapy for kinésie paradoxale. Neurol Med (Tokyo) 6:151–156

Lakke PWF (Chairman of ad hoc Committee on Classification) (1981) Classification of extrapyramidal disorders. J Neurol Sci 51:311–327

Martin JP (1967) The basal ganglia and posture. Pitman Medical, London

Nagasaki H, Kosaka K, Nakamura R (1981) Disturbances of rhythm formation in patients with hemiplegic lesion. Tohoku J Exp Med 135:231–236

Nakagawa T, Uyama E, Kumamoto T, Uchino M, Araki S (1987) A case of pure akinesia with pheochromocytoma. Clin Neurol (Tokyo) 27:1150–1153

Nakamura R, Nagasaki H, Narabayashi H (1976) Arrhythmokinesia in parkinsonism. In: Birkmayer W, Hornykiewicz O (eds) Advances in parkinsonism. Roche, Basle, pp 258–268

Nakane K, Wakai S, Ishikawa Y, Nagaoka M, Minami R, Shinoda M (1986) Kinésie paradoxale in a case of Wilson's disease. Neurol Med (Tokyo) 25:25–30

Narabayashi H, Imai H, Yokochi M, Hirayama K, Nakamura R (1976) Cases of pure akinesia without rigidity and tremor and with no effect by L-dopa therapy. In: Birkmayer W, Hornykiwicz O (eds) Advances in parkinsonism. Roche, Basle, pp 335–342

Narabayashi H, Kondo T, Hayashi A, Suzuki T, Nagatsu T (1981) L-threo-3,4-dihydroxyphenylserine treatment for akinesia and freezing of parkinsonism. Proc Jpn Acad 57 [SerB]:351–354

Narabayashi H, Kondo T, Yokochi F, Nagatsu T (1986) Clinical effect of L-threo-3,4-dihydroxyphenylserine in cases of parkinsonism and pure akinesia. Adv Neurol 45:593–602

Parkinson J (1817) An essay on the shaking palsy. Sherwood, Neely & Jones, London

Quinn NP, Luthert P, Honavar M, Marsden CD (1989) Pure akinesia due to Lewy body Parkinson's disease: a case with pathology. Movement Disorders 4:85–89

Schwab RS, England AC (1968) Parkinson syndromes due to various specific causes. In: Vinken PJ, Bruyn GW (eds) Handbook of clinical neurology, vol 6. Disease of the basal ganglia. North-Holland, Amsterdam, pp 227–247

Schwab RS, Zieper I (1965) Effects of mood, motivation, stress and alertness on the performance in Parkinson's disease. Psychiat Neurol (Basel) 150:345–357

Senoh Y, Hirai S, Morimatsu K, Okamoto K, Isaka K (1982) Kinésie paradoxale. Analysis of two clinical cases with different causes and lesions. Neurol Med (Tokyo) 16:260–270

Souques MA (1921) Rapport sur les syndromes parkinsoniens. Rev Neurol 28:534–573

Takahashi K, Nakashima R, Takao T, Nakamura H (1977) Pallido-nigro-luysial atrophy associated with degeneration of the centrum medianum. Acta Neuropathol (Berl) 37:81–85

Tokiguchi S, Natsukawa M, Kawase Y, Hayashi H, Tsukada Y (1978) Pure akinesia with apraxia of lid-opening. Clin Neurol (Tokyo) 18:208–212

Yamamoto M, Ujike H, Ogawa N (1985) Effective treatment of pure akinesia with L-threo-3,4-dihydroxyphenylserine (DOPS): report of a case, with pharmacological considerations. Clin Neuropharmacol 8:334–342

Yuasa T, Homma Y, Takahashi H, Mori S, Hayashi H (1987) Progressive supranuclear palsy with pure akinesia as an initial symptom. Neurol Med (Tokyo) 26:460–467

Correspondence: Dr. H. Imai, Department of Neurology, Juntendo University School of Medicine, 2-1-1 Hongo, Bunkyo-ku, Tokyo 113, Japan.

Hereditary progressive dystonia with marked diurnal fluctuation

M. Segawa and Y. Nomura

Segawa Neurological Clinic for Children, Tokyo, Japan

Summary

Hereditary progressive dystonia with marked diurnal fluctuation is a particular levodopa-responsive dystonia with onset in childhood. Clinically, it is characterized by generalized postural dystonia which becomes aggravated towards evening and is markedly alleviated in the morning after sleep. There was no axial torsion or action dystonia observed. In later life, it is associated with postural tremor, not parkinsonian resting tremor. The freezing phenomenon or petit pas is not present, and locomotive activity is maintained even in advanced stages. Although symptoms progress in the first 3 decades of life, they become static afterwards. Levodopa shows marked and sustained effects without unfavorable side effects. Autosomal dominant inheritance with low penetrance is suspected, but there is marked female predominance. The main lesion is thought to be a hypofunctioning of tyrosine hydroxylase in the terminal of the nigrostriatal dopamine neuron which mainly affects the direct striatal projections to the pars reticulata of the substantia nigra and the medial segment of the globus pallidus.

Introduction

Hereditary progressive dystonia with marked diurnal fluctuation (HPD) is characterized by aggravation of symptoms towards evening and their alleviation in the morning after sleep. Since our first report (Segawa et al., 1972) more than 70 cases have been reported from Japan and other countries, including our personal cases. In this paper we review the characteristics of HPD and demarcate this disorder in basal ganglia diseases.

Clinical characteristics of our personal cases

Our 18 cases include 12 familial cases from 6 families and 6 sporadic cases. There is marked sex preference for females (total: F:M = 15:3; familial cases 11:1, sporadic cases 4:2). This disorder develops insidiously in a normal child without any episodes of causing chronic dopaminergic dysfunction.

Onset of the clinical symptoms ranges from 1 year to 11 years of age with an average at 6.03 ± 2.82 years. There is a slight difference between averages of familial (6.12 ± 2.97 years) and sporadic cases (5.85 ± 2.73 years) and between female (6.10 ± 2.91 years) and male cases (5.60 ± 2.78 years). The initial symptoms in most cases are fatigability and gait disturbance due to leg dystonia, with flexion-inversion of one foot (pes equinovarus). However, cases with later onset may present dystonia in one arm with flexion-pronation of the elbow and palmar-flexion of the wrist and may sometimes be associated with postural tremor. These symptoms are aggravated towards the envening and are markedly alleviated or completely disappear in the morning after sleep.

The symptoms progress gradually, spread to other extremities and within 5 or 6 years, all limbs become involved. Some cases show the orderly sequence following the shape of letter "N" (Segawa et al., 1986a).

The basic symptom is postural dystonia which becomes apparent when a sitting or standing position is adopted, and is aggravated by voluntary movements or psychological tension. When the patient is standing, the posture appears to be similar to spastic paraplegia with lumbar lordosis, but in some cases, flexor posture of the trunk, similar to "pallidal posture," develops when the patient is sitting on the floor or standing with support. However, no cases show action dystonia or axial torsion. Neck muscle involvement is mild with minimal retrocollis; no cases show torticollis.

Neurological examinations reveal rigidity in the affected muscles. In the initial stage, rigidity appears depending on the postures and movements. In the advanced stage, it is observed constantly, but is not plastic; thus by repeating stretch reflexes, rigid hypertonus can be occasionally abolished and proves to be hypotonic. The Westphal phenomenon is observed in most cases, particularly in the advanced stage. Postural tremor may become apparent after 10 years of age or later, and in some cases, it appears in the 4th decade of life (Segawa et al., 1986a, 1990). However, no case shows parkinsonian resting tremor or intension tremor. The pseudo-Babinski posture is observed occasionally, depending on the posture and voluntary movements that are initiated. Dysdiadochokinesis is observed in all cases, but finger-nose and knee-heel tests are normal on most cases.

The gait is rigid with marked dystonic posture of the extremities, and lumbar lordosis is present. Coordinated movements of the arms are impaired in the affected side in the early stage, but they disappear on both sides in advanced stages. However, no cases show claw-hand and claw-toe posture or petit pas of parkinsonism. No cases show difficulty in crawling. Although in the advanced stage, there is flexor posture in the legs, the pattern of crawling is normal with swing phase initiated by the legs (Segawa and Nomura, 1991a). Slowness in movement is observed in all cases and later becomes bradykinetic. Pulsion is observed in advanced cases, but it is mild, and none of the cases shows the freezing phenomenon. The face is expressionless and later becomes masklike; in advanced cases, dysarthria may be observed.

The deep tendon reflexes are exaggerated, particularly in the legs. In some, ankle clonus is observed. The pseudo-Babinski reaction, striatal foot, is also detected, but the plantar response is flexor. The tilting reaction is affected, particularly on the side with more affected limbs. There is no restriction in the ocular movements: the optic fundi are normal, and there is no Kayser-Fleischer ring present. There are no abnormalities in the cerebellum, in the sensory system or in the mental and psychological activities.

The main clinical feature is dystonia which persists throughout the course of the illness. In advanced stages, the symptoms are more marked in the legs, even in cases whose initial symptoms were in the arms. At severely advanced stages, pes equinovarus contracture may develop. The asymmetry, with left-side pre-dominance, is also observed throughout the course of the disease (L:R = 11:7). The diurnal fluctuation becomes less evident with the progression of the diseases, when marked generalized dystonia is observed even in the moring.

In a case (S.S.) with a clinical course of 43 years without medical treatment, walking became impossible around 15 years of age, 8 years after the onset, but from the third decade, the progression of symptoms became less apparent. After the 4th decade, it became almost static with attenuation of the diurnal fluctuation, and postural tremor appeared (Segawa et al., 1986a,b, 1990).

The whole clinical course can be observed even in the early stage on the diurnal fluctuation of symptoms. However, the postural tremor does not appear even in the evening at younger ages. The symptoms are milder, or not complete, and the progression is slow in cases with clinical onset after ten years of age. Affected parents also have a milder course than their children.

Cases reported

Up to now, more than 70 cases have been reported in Japan and other countries under the name of HPD or levodopa-responsive dystonia. However, among them there are cases different from HPD, either in symptoms or in the pattern of response to levodopa (see below). Excluding these, 44 remain as typical cases, 20 from Japan and 24 from other countries. Most of the cases showed initiations of symptoms with leg dystonia of one leg in the first decade of life (Japanese: 6.2 ± 2.0 years, other countries: 4.7 ± 2.4 years). They showed female predominance (Japanese: 16:4, other countries: 18:6) and side preference to the left in cases where laterality of the symptoms is described (Japanese 13:5, other countries 4:2).

Most Japanese cases showed shorter body length than average, which became manifest with the onset of dystonia, but increased to normal range after levodopa (Segawa et al., 1976; Shimizu et al., 1986).

Investigation

Surface EMG reveals nonreciprocal tonic involuntary activities. In aged cases, grouping discharges and tonic stretch reflexes are observed, often with the Westphal phenomenon.

Polysomnography (PSG) revealed abnormalities in two kinds of body movements during sleep (Segawa et al., 1976, 1986a,b; Hakamada et al., 1982; Nomura et al., 1987b); gross movements (GMs) — diffuse sequential muscle activities including those of the rectus abdominalis, lasting more than 2 seconds — showed abnormality in the sleep-stage-dependent modulation with an increase of their rate in stage 2, a decrease in stage 1 and sREM; and twitch movements (TMs) — short EMG activities localized to one muscle, lasting less than 0.5 seconds — decreased in number in all sleep stages but the sleep-stage-dependent modulation was preserved normally. In contrast, the number of rapid eye movements (REMs) increased. Therefore, the ratio of the number of TMs of the mentalis muscle during the REM stage against that of REMs (ment TM sREM/REMs) was below the normal range. There is a directional preference in horizontal REMs towards the left side in contrast to the normal preference towards the right (Segawa et al., 1991b). These abnormalities are normalized after levodopa. During sleep, the number of TMs increased and the pattern of rolling over gradually normalized when the later sleep cycles were approached (Segawa, 1982). This shows the existence of a nocturnal variation of symptoms in this disorder in addition to a diurnal fluctuation. Selective sleep deprivation of stage 4 or REM, performed in two cases, revealed that the REM stage has an important role in the ameliorative effect of sleep (Segawa et al., 1976).

Examination of voluntary saccade in HPD revealed slow and hypometric anticipatory saccade, similar to Parkinson's disease (Nomura et al., 1987a).

Estimation of catecholamine metabolites in CSF revealed low levels of HVA (Kumamoto et al., 1984; Ouvrier, 1978; Shimoyamada et al., 1986), and they were lower in the morning than in the afternoon (Kumamoto et al., 1984; Shimoyamada et al., 1986).

Brain CT scan and MRI were normal in our group. PET scans revealed decreased absorption of 18 Fluodopa in the left putamen of an 11-year-old dominantly inherited case, but those of a 34-year-old sporadic case were normal (Lang et al., 1988). No abnormalities were observed in the levels of serum ceruloplasmin, copper, copper excretion in urine or serum levels of pyruvate and lactic acid; the routine laboratory examination results were normal.

Treatment

In most cases, a dose of 20 mg/kg/day of levodopa without inhibitor alleviates symptoms completely. In a few cases, choreic movements were developed by a rapid increase of dosage or by erronous administration of a double dose of levodopa, but these symptoms were quickly eliminated by reducing the dose (Segawa et al., 1990). In triplet cases, early levodopa showed better effects (Tachi et al., 1987). However, the studies of 8 cases who had been administrated with levodopa for more than 10 years (Segawa et al., 1990) revealed that in HPD, the effects of levodopa persist without unfavorable side effects and that the dosage could later be reduced. Case S.S., who was started on levodopa at the age of 51 years, became ambulant within 3 days under 600 mg levodopa after a 36-year inambulant period (Segawa et al., 1990). She is now 70 years of age and is completely normal under a 50 to 100 mg menesit treatment.

However, in cases who were started on levodopa in childhood, periods of subjective feeling of ineffectiveness developed, and further increase in dosage was necessary. These feelings occurred in the leg in which the first symptoms occurred, but were without objective signs except in one case. This feature occurred from 2 months to 8.6 years (mean 4.2 ± 3.0 years) after the start of levodopa treatment, but this period was age dependent, and showed development at ages from 11.7 to 14.7 years (mean 12.8 ± 1.2 years). This feature was not observed in those whose levodopa treatments were started after the mid-teens. This feeling of ineffectiveness was alleviated by increasing the dosage of levodopa up to 30 mg/kg/day, and in some cases by additional administration of a decarboxylase inhibitor. PSG performed during the period of this feeling revealed a decrease in the number of TMs in the muscle concerned. However, there were no abnormalities in the other parameters including the sleep-stage-dependent modulation of GMs (Segawa et al., 1990).

In HPD, the time course of plasma DOPA after administration of levodopa (without decarboxylase inhibitor) showed marked variation according to age; i.e., cases examined under 12 years of age revealed low levels, while in those over 12 years, the levels increased significantly, even with the same doses of levodopa. Cases with feelings of ineffectiveness showed the pattern of childhood even in adolescence (Segawa et al., 1990).

The anticholinergic drug had a marked and prolonged effect, but did not afford complete relief, neither clinically nor polysomnographically, and complete recovery was obtained after replacing it with levodopa (Segawa et al., 1990). Bromocriptine was also effective but did not afford complete relief (Nomura et al., 1987b). Tetrahydrobiopterin (BH4) treatment was attempted in HPD patients (Ishida et al., 1988; LeWitt et al., 1976, 1983), but thus far, in no cases of HPD has BH4 shown more favorable effects than levodopa, except for in one case of Ishida et al. (1988).

Inheritance

We speculate that HPD comes from autosomal dominant inheritance with low penetrance (Segawa et al., 1976), but recessive inheritance is also suspected (Ouvrier, 1978; Kumamoto et al., 1984; Deonna, 1986). Thus, no substantial data exist yet for deciding the mode of inheritance. However, cases suggesting recessive inheritance have different clinical features than HPD and may belong to a different category (see below).

Differential diagnosis

HPD is differentiated from idopathic torsion dystonia by an absence of axial torsion and marked response to levodopa. This differentiation is due to whether the lesion is in the nigrostriatal (NS) dopamine (DA) neuron (HPD) (Segawa et al., 1988) or in the basal ganglia (torsion dystonia) (Rothwell and Obeso, 1988). Cases with action dystonia or oculogyric crisis differ from HPD by having postsynaptic supersensitivity at the DA receptor of the basal ganglia revealed by PSG (Segawa et al., 1987, 1988).

Not all cases of levodopa-responsive dystonia have diurnal fluctuation of symptoms (Deonna, 1986; Nygaard and Duvoison, 1986; Nygaard et al., 1988). Of those cases without diurnal fluctuation, some may be HPD because of having family members with typical HPD or having other characteristics of HPD. However, others might be different because of an incomplete response to levodopa with or without unfavorable side effects; action dystonia, resting tremor, oculogyric crisis or early bulbar involvement (Segawa, 1989). Among cases with diurnal fluctuation, those with oculogyric crisis, torticollis or on-and-off phenomenon might be different from HPD (Segawa, 1989). Cases 3 and 4 of Fink et al.'s report (1988) might be biopterine deficiency because of incomplete response to levodopa and steady progressive courses (Segawa, 1989).

Levodopa-responsive dystonia with features different from HPD tends to have affected siblings, suggesting recessive inheritance, while those having identical or almost the same features as HPD are transgeneration cases.

Among cases reported as juvenile Parkinson's disease or parkinsonism (JPA) or Parkinson-dystonia complex, some developed symptoms before 10 years of age (Allen and Knopp, 1976) or had diurnal fluctuation of symptoms (Yamamura et al., 1973; Sunohara et al., 1985). However, they differ from HPD with parkinsonian resting tremor, movement related fatigability (Sunohara et al., 1985) and progressive course with decrease in responsiveness to levodopa (Yamamura et al., 1973, personal communication).

Yokochi (1979) and Narabayashi et al. (1986) classified JPA into two types: one having clinical symptoms similar to classical idiopathic Parkinson's disease and the other with dystonia as a main symptom throughout the course of the illness. Recently, Quinn et al. (1987) documented the same results. The author (MS) termed the former group tremor-type JPA and the latter dystonic-type JPA, and demonstrated their difference from HPD (Segawa, 1981; Segawa et al., 1986a,b).

Compared to Yokochi's studies (Yokochi, 1979), tremor-type JPA is easily differentiated from HPD by age of onset (29.65 ± 6.9 years), parkinsonian resting tremor, and easier development of up-and-down phenomenon and dyskinesia after levodopa. This type of JPA shows no sex preference, marked pulsion with freezing and difficulty in crawling with total flexor posture (Segawa et al., 1991a). Dystonic-type JPA [group III of Yokochi (1979)] has the age of onset as 11.9 ± 4.3 years; slightly but significantly older than that of HPD and differing from HPD in male predominance (F:M = 1:5) with a tendency toward up-and-down phenomenon and dyskinesia after levodopa.

Dystonia caused by biopterine deficiency differs from HPD by a relatively poor response to levodopa and necessitates addition of biopterine. Clinically, these cases show symmetrical involvement with lack of interlimb coordination. However Ishida et al.'s case (1988) suggests involvement of serotonergic systems other than dopamine or the role of biopterine in HPD.

There are families in which HPD coexists with torsion dystonia with poor response to levodopa (Sterk and Kellermann, 1984), parkinsonism with onset in the 4th decade of life (Nomoto et al., 1983) and type III of Yokochi's JPA (Ujike et al., 1989). The latter might be JPA because it suggests recessive inheritance. Nomoto et al.'s (1983) parkinsonism showed mild dystonia and postural tremor which responded markedly to relatively low doses of levodopa and were thought to be mild or abortive late-onset HPD. Since late-onset cases of HPD are always mild and abortive, the Sterk and Kellermann cases (1984) are quite exceptional as they suggest a combination of different pathophysiologies.

Pathophysiology and possible pathology

The dramatic and sustained response to levodopa without unfavorable side effects implicates HPD as a functional disorder in which the lesions are restricted to the nigrostriatal DA neuron. Early-onset extrapyramidal disorders appear as dystonia. Onset of dystonia in the first decade on life generally tends to begin with leg dystonia (Marsden and Harrison, 1974). The pathophysiology of HPD should be considered with reference to the age variation of the dopamine system.

The age of onset of HPD corresponds to the period of marked aged variation of the striatal tyrosine hydroxylase (TH) activities (McGeer and McGeer, 1973; McGeer et al., 1977). The clinical course of HPD, i.e., marked progression in childhood with later attenuation in the 3rd decade of life and almost static after the 4th decade just follows the age-related decline curve of the striatal TH activities. Marked diurnal fluctuation (aggravation) as well as nocturnal fluctuation (allevication) of symptoms is thought to reflect the circadian oscillation of the striatal TH activities observed in the rat brain (McGeer and McGeer, 1973).

It is suggested then that the main lesion of HPD is at the terminal of the NS-DA neuron with a decrease in TH activities (Segawa, 1981; Segawa et al., 1986a). A similar pathophysiology is suspected in dystonic-type JPA, while in the tremor-type JPA, the main lesion may be in the substantia nigra (Segawa, 1981; Segawa et al., 1986a) since its age of onset corresponds to the period of cell loss in the structure (McGeer and McGeer, 1973). Two autopsied cases, one dystonic- and the other tremor-type JPA (Yokochi et al., 1984), support this consideration.

The basal ganglia modulate locomotion via the pedunculo-pontine-nuclei (PPN) with two efferents: one from the substantia nigra pars reticulata (SNr) and the other from the medial segment of the globus pallidus (MGP) (Garcia-Rill, 1986). For development of levodopa-induced dyskinesia, an indirect projection of the striatum, i.e., putamen-lateral segment of the globus pallidus (LGP)-subthalamic nucleus (STN)-MGP, which is connected to the thalamic nuclei, is thought to play important roles (Crossman, 1990). For voluntary saccadic movement, the caudate nucleus (CN)-SNr-superior colliculus pathway is involved (Hikosaka and Sakamoto, 1986) and the same pathway seems to play a role for modulation of REMs during the REM stage (Segawa et al., 1986b, 1987, 1991b). GMs during sleep are thought to be modulated by the pallidofugal thalamic pathway from MGP and TMs by the striatofugal descending pathway, probably via SNr (Segawa et al., 1987).

Correlation of the clinical and laboratory findings of HPD to this evidence implicates that in HPD, the DA terminal connecting to the striatal efferent to SNr is mainly involved, and those connecting to other direct efferents to the MGP are also affected, but only mildly, while those connected to the indirect pathway might be preserved. These intrastriatal structures per se should be preserved, suggested by the lack of axial torsion, marked and sustained responsiveness to levodopa and preservation of dopa-acetylcholine interaction observed in polysomnographic examination (Segawa et al., 1976, 1986a, 1987). Postsynaptic supersensitivity might not exist because of the absence of action dystonia or of oculogyric crisis (Segawa et al., 1987).

The importance of the direct pathway in dystonia is shown in the neuropathological examination of rigid-type Huntington's disease (Albin et al., 1990). Sparing of the pathways connecting to STN in dystonia is also suggested when comparing the deoxyglucose cooperation of the basal ganglia between levodopa-induced dystonia and dyskinesia in MTPT monkeys (Crossman, 1990a).

Side preference might be a reflection of physiologic asymmetry in the activities of the basal ganglia observed in human autopsies (Glick et al., 1982). The abnormal direction of REMs with leftward preference (Segawa et al., 1986a, 1991b) also implies that the asymmetry of symptoms depends on the functional asymmetry of NS-DA neurons.

Female predominance of HPD, in contrast to male predominance of dystonic-type JPA, may prove to be pathognomonic, when the reverse sex preferences observed in disorders with exaggerated DA transmission are compared (i.e., male predominance of Gilles de la Tourette syndrome with onset in childhood and female predominance of Syndenham chorea with onset in adolescence). This evidence, with the age-related alteration in the time course of plasma DOPA after oral levodopa, suggests alteration of modulation of the striatum DA activity between the first and second decades of life.

The mode of inheritance of HPD is suggested as being dominant with low penetrance. This implies that enzyme deficiency or the degeneration process is unlikely to be the pathogenesis. The gene study of the locus of tyrosine hydroxylase showed negative results.

HPD is a particular disorder different from JPA or Parkinson's disease. Among levodopa-responsive dystonias, this disease is demarcated as a dominantly inherited type. Clarification of the pathophysiology requires further evidence from neuropathological and neurohistochemical studies.

References

Albin RL, Reiner A, Anderson KD, Penney JB, Young AB (1990) Striatal and nigral neuron subpopulations in rigid Huntington's disease: implications for the functional anatomy of chorea and rigidity-akinesia. Ann Neurol 27:357–367

Allen N, Knopp W (1976) Hereditary parkinsonism-dystonia with sustained control by L-Dopa and anticholinergic medication. In: Eldridge R, Fahn S (eds) Advances in neurology, vol 14. Raven Press, New York, pp 201–213

Crossman AR (1990) A hypothesis on the pathophysiological mechanisms that underlie levodopa dopamine against induced dyskinesia in Parkinson's disease. Implications for future strategies in treatment. Movement Disorders 5:100–108

Crossman AR (1990a) Animal models of movement disorders. Paper presented at the First International Congress of Movement Disorders. Washington, D.C.

Deonna T (1986) Dopa-sensitive progressive dystonia of childhood with fluctuations of symptoms – Segawa's Syndrome and possible variants. Neuropediatrics 17:75–80

Fink JK, Barton N, Cohen W, Lovenberg W, Burns RS, Hallett M (1988) Dystonia with marked diurnal variation associated with biopterin deficiency. Neurology 38:707–711

Garcia-Rill E (1986) The basal ganglia and the locomotor regions. Brain Res Rev 11:47–63

Glick SD, Ross DA, Hough LB (1982) Lateral asymmetry of neurotransmitters in human brain. Brain Res 234:53–63

Hakamada S, Watanabe K Hara, K, Miyazaki S (1982) A case of "hereditary progressive dystonia with marked diurnal fluctuation." No To Hattatsu (Tokyo) 14:44–48

Hikosaka O, Sakamoto M (1986) Cell activity in monkey caudate neuclus preceding saccadic eye movements. Exp Brain Res 63:619–622

Ishida A, Takada G, Kobayashi Y, Higashi O, Toyoshima I, Takai K (1988) Involvement of serotonergic neuron in hereditary progressive dystonia – Clinical effects of tetrohydrobiopterine and 5-hydroxytryptophan. No To Hattatsu (Tokyo) 20: 195–199

Kumamoto I, Nomoto M, Yoshidome M, Osame M, Igata A (1984) Five cases of dystonia with marked diurnal fluctuation and special reference to homovanillic acid in CSF. Clin Neurol (Tokyo) 24:697–702

Lang AE, Garnett ES, Firnau G, Nahamaias C, Talalla A (1988) Positron tomography in dystonia. In: Fahn F, Marsden CD, Calne DB (eds) Advances in neurology, vol 50. Dystonia 2. Raven Press, New York, pp 249–253

LeWitt PA, Miller LP, Newman RP, Lovenberg W, Eldridge R, Chase TN (1976) Pteridine cofactor in dystonia: pathogenic and therapeutic considerations. Neurology (2) 33:161

LeWitt PA, Newman RP, Miller LP, Lovenberg W, Eldridge R (1983) Treatment of dystonia with tetrahydrobiopterin. N Engl J Med 308:157–158

Marsden CD, Harrison MHG (1974) Idiopathic torsion dystonia (dystonia musculorum deformans): a review of 42 patients. Brain 97:793–810

McGeer EG, McGeer PL (1973) Some characteristics of brain tyrosin hydroxylase. In: Mandel J (ed) New concepts in neurotransmitter regulation. Plenum Press, New York London, pp 53–68

McGeer PL, McGeer EG, Suzuki JS (1977) Aging and extrapyramidal function. Arch Neurol 34:33–35

Narabayashi H, Yokochi M, Iizuka R, Nagatsu T (1986) Juvenile parkinsonism. In: Vinken PJ, Bruyn GW, Klawans HL (eds) Handbook of clinical neurology, vol 5 (49). Extrapyramidal disorders. Elsevier Science Publishers, Amsterdam, pp 153–165

Nomoto M, Kumamoto K, Sano Y, Nakajima H, Osame M, Igata A (1983) A family of benign juvenile parkinson disease with a child having severe dopa-responsive fluctuating dystonia. Clin Neurol (Tokyo) 24:1388

Nomura Y, Segawa M, Soda M, Hikosaka O (1987a) Voluntary saccadic eye movements in basal ganglia disorders. In: Highlights in Neuro-Ophthalmology — Proceedings of the Sixth Meeting of the International Neuro-Ophthalmology Society, Hakone, Japan, 1986. Aeolus Press, Amsterdam, pp 139–145

Nomura K, Negoro T, Takasu E, Aso K, Furune S, Takahashi I, Yamamoto N, Watanabe K (1987b) Bromocriptine therapy in a case of hereditary progressive dystonia with marked diurnal fluctuation. Brain Dev 9:199

Nygaard TG, Duvoison RC (1986) Hereditary dystonia-parkinsonism syndrome of juvenile onset. Neurology 36:1424–1428

Nygaard TG, Marsden CD, Duvoisin RC (1988) Dopa-responsive dystonia. In: Fahn S, Marsden CD, Calne DB (eds) Advances in neurology, vol 50. Dystonia 2. Raven Press, New York, pp 377–384

Ouvrier RA (1978) Progressive dystonia with marked diurnal fluctuation. Ann Neurol 4:412–417

Quinn N, Critchley P, Marsden CD (1987) Young onset Parkinson's disease. Movement Disorders 2:73–91

Rothwell JC, Obeso JA (1988) The anatomical and physiological basis of torsion dystonia. In: Marsden CD, Fahn S (eds) Movement disorders, neurology 2. Butterworth, London, pp 367–376

Segawa M. (1981) Hereditary progressive dystonia (HPD) with marked diurnal fluctuation. Adv Neurol Sci (Tokyo) 25:73–81

Segawa M (1982) Catecholamine metabolism in neurological diseases in childhood. In: Wise G, Blaw ME, Procopis PG (eds) Topics in child neurology, vol 2. Spectrum Publications, New York, pp 135–150

Segawa M (1989) Hereditary progressive dystonia with marked diurnal fluctuation. Pharma Medica (Jpn) 7:171–183

Segawa M, Hosaka A, Miyagawa F, Nomura Y, Imai H (1976) Hereditary progressive dystonia with marked diurnal fluctuation. In: Eldridge R, Fahn S (eds) Advances in neurology, vol 14. Raven Press, New York, pp 215–233

Segawa M, Nomura Y (1991a) Pathophysiology of human locomotion: studies on clinical cases. In: Shimamura M, Grillner S, Edgerton RV (eds) Neurobiological basis of human locomotion. Japan Scientific Societies Press, Tokyo, pp 317–328

Segawa M, Nomura Y (1991b) Rapid eye movements during stage REM are modulated by nigrostriatal dopamine (NS-DA) neurons? In: Bernardi G, Carpenter MB, Di Chiara G (eds) Basal ganglia III (in press)

Segawa M, Nomura Y, Hikosaka O, Soda M, Usui S, Kase M (1987) Roles of the basal ganglia and related structures in symptoms of dystonia. In: Carpenter MB, Jayaraman A (eds) Basal ganglia II — structure and function. Plenum Press, New York, pp 489–504

Segawa M, Nomura Y, Kase M (1986a) Diurnally fluctuating hereditary progressive dystonia. In: Vinken PJ, Bruyn GW, Klawans HL (eds) Handbook of clinical neurology, vol 5 (49). Extrapyramidal disorders. Elsevier Science Publishers, Amsterdam, pp 529–539

Segawa M, Nomura Y, Kase M (1986b) Hereditary progressive dystonia with marked diurnal fluctuation: clinicopathophysiolgical identification in reference to juvenile Parkinson's disease. In: Yahr MD, Bergmann KJ (eds) Advances in neurology, vol 45. Parkinson's disease. Raven Press, New York, pp 227–234

Segawa M, Nomura Y, Tanaka S, Hakamada S, Nagata, E, Soda M, Kase M (1988) Hereditary progressive dystonia with marked diurnal fluctuation: consideration on its pathophysiology based on the characteristics of clinical and polysomnographical findings. In: Fahn F, Marsden CD, Calne DB (eds) Advances in neurology, vol 50. Dystonia 2. Raven Press, New York, pp 367–376

Segawa M, Nomura Y, Yamashita S, Kase M, Nishiyama N, Yukishita S, Ohta H, Nagata K, Hosaka A (1990) Long term effects of L-Dopa on hereditary progressive dystonia with marked diurnal fluctuation. In: Berardelli A, Benecke R, Manfredi M, Marsden CD (eds) Motor disturbances II. Academic Press, London, pp 305–318

Segawa M, Ohmi K, Itoh S, Aoyama M, Hayakawa H (1972) Childhood basal ganglia disease with remarkable response to L-Dopa, "hereditary basal ganglia disease with marked diurnal fluctuation." Shinryo (Tokyo) 24:667–672

Shimizu N, Hara M, Yoshihara S, Tateno A, Aoki T (1986) A familized (mother-son cases) hereditary progressive dystonia with marked diurnal fluctuation. Shonika Rinsho (Tokyo) 39:1442–1446

Shimoyamada Y, Yoshikawa A, Kashio M, Kihira S, Koike M (1986) Hereditary progressive dystonia – an observation of the catecholamine metabolism during L-Dopa therapy in a 9-year-old girl. No To Hattatsu (Tokyo) 18:505–509

Sterk E, Kellerman K (1984) Unterschiedliche Expressivität des Segawa-Syndroms in einer Familie. Paper presented at the 10th Annual Meeting of the "Gesellschaft für Neuropädiatrie", October 5–7, 1984, Giessen, Germany

Sunohara N, Mano Y, Ando K, Satoyoshi E (1985) Idiopathic dystonia: parkinsonism with marked diurnal fluctuation of symptoms. Ann Neurol 17:39–45

Tachi N, Sasaki K, Shinoda M (1987) Four cases including identical triplets of progressive dystonia with marked fluctuation. J Jpn Pediat Soc 91:1403–1406

Ujike H, Nakashima M, Kuroda S, Otsuki S (1989) Two siblings of juvenile Parkinson's disease dystonic type (Yokochi type 3) and hereditary progressive dystonia with marked diurnal fluctuation (Segawa). Clin Neurol (Tokyo) 29:890–894

Yamamura Y, Sobue I, Ando K, Iida M, Yanagi T, Kono C (1973) Paralysis agitans of early onset with marked diurnal fluctuation of symptoms. Neurology 23:239–244

Yokochi M (1979) Juvenile Parkinson's disease, part 1. Clinical aspects. Adv Neurol Sci (Tokyo) 23:1060–1073

Yokochi M, Narabayashi H, Iizuka R, Nagatsu T (1984) Juvenile parkinsonism – some clinical, pharmacological and neuropathological aspects. In: Hassler RG, Christ JF (eds) Advances in neurology, vol 40. Parkinson's disease. Raven Press, New York, pp 407–413

Correspondence: Dr. M. Segawa, Segawa Neurological Clinic for Children, Segawa Building 2 Fl., 2–8 Surugadai Kanda, Chiyodaku, Tokyo 101, Japan.

Positron emission tomographic study in Parkinson's disease—rigid vs tremor type

C. Ohye with the collaboration of T. Shibazaki, M. Hirato,
Y. Kawashima, M. Matsumura, S. Horikoshi, and T. Shibasaki

Department of Neurosurgery, Gunma University School of Medicine, Gunma, Japan

Summary

Using ^{15}O (O_2, CO_2) and ^{18}F-fluoro-deoxyglucose, a positron emission tomographic study was carried out to elucidate ongoing changes in a whole brain in cases with rigid- and tremor-type Parkinson's disease (PA), respectively. For comparison, cases with essential tremor and normal control were also examined. MRI and XCT images and depth recording from basal ganglia during the course of stereotactic thalamotomy were considered together.

In rigid-type PA, cortical rCBF, $rCMRO_2$ and rCMRGl were significantly reduced. However, ^{18}FDG in basal ganglia (especially in the region of putamen and pallidum) was maintained in the normal or even in higher level. Microrecording also revealed reduced electrical activity in caudate nucleus, in contrast with higher activity in pallidum.

In tremor-type PA, the reduction of cortical metabolic rate was moderate, and activity in the caudate was maintained.

Metabolic rate in essential tremor and normal case seemed to be similar: it was high in cortical level and basasl ganglia (Cd was more active than putamen-pallidum).

Different metabolic features between rigid- and tremor- type PA were emphasized.

Introduction

Positron emission tomography (PET) can demonstrate the functional state of the living human brain. Thus, specific functional brain mapping has been made during speech, memory task, visual or auditory activity, mostly in normal volunteers. Also, PET studies on abnormal brain of organic and functional disorder are in progress. Among the functional disorders, epilepsy, involuntary movement and intractable pain are the main subjects of PET study (Brooks et al., 1989; Martin, 1985; Phelps et al., 1985; Shibasaki, 1989; Wagner, 1986). At present, however, in Parkinson's disease, the reported results were not consistent. For example, the oxygen or glucose metabolic rate of the basal ganglia was claimed to be either normal (Leenders et al., 1985), increased

(Martin, 1985; Raichle et al., 1984; Rougemont et al., 1984; Eolfson et al., 1985) or even decreased (Eidelberg et al., 1990; Kuhl et al., 1984).

We are interested especially in the neural mechanism of involuntary movement, and PET scan is used in combination with stereotactic depth recording, and other computerized imaging such as XCT and MRI, in order to get more precise information about ongoing changes in such a disordered brain as a whole (Hirato et al., 1989; Ohye, 1985, 1989, 1990; Ohye et al., 1990). In this study, we have selected rigid- and tremor-type Parkinsonism respectively and considered them separately.

Subjects and methods

Three cases of rigid-type Parkinson's disease, four cases of tremor-type Parkinson's disease and five cases of essential tremor were studied (Table 1). All cases with Parkinson's disease had been treated elsewhere by neurologists (see Table 1), and finally was sent to us for surgical treatment because drug therapy was no longer effective. All underwent PET study (Hitachi PET-H1) and then stereotactic thalamotomy with microrecording. As a control, two normal persons had PET study, without surgery, Using ^{15}O, O_2 or CO_2 gas was inhaled (10 mci/min) and steady state measurement of regional cerebral blood flow (rCBF) and regional cerebral oxygen metabolic rate (rCMRO$_2$) was performed.

In recent cases, 18F-FDG was used to measure the regional cerebral glucose metabolic rate (rCMRG1). Five mCi of FDG was injected intravenously and measurement started 60 min after bolus injection. For quantitative analysis,

Table 1. Patient list

Diagnosis	Case	Age	Sex	Side	Duration	L-Dopa
PA	K.S.	64	F	rt	2y	(−)
(h-tremor)	K.H.	63	M	rt	8y	(−)
	K.H.	52	F	rt	7y	(−)
	K.K.	70	M	lt	5y	2y
PA	K.H.	61	F	(rt)	6y	6y
(rigid)	K.M.	55	M	(rt)	12y	10y
	Y.O.	42	M	rt	8y	6y
Ess-T	M.I.	53	M	rt	1y	(−)
	T.K.	51	M	rt	(3y)	(−)
	M.O.	31	M	rt > lt	12y	(−)
	N.O.	53	M	rt > lt	8y	(−)
	S.K.	66	M	lt > rt	10y	(−)
Normal	Y.S.	50	M	/	/	(−)
	M.K.	28	M	/	/	(−)

PA Parkinson's disease; *Ess-T* Essential tremor; *h-tremor* hemi-tremor; *h-rigid* hemi-rigidity

blood sampling was performed 30 times for 60 min after injection of FDG and the radioactivity was counted. The metabolic rate was calculated using the kinetic rate constant proposed by Phelps et al. (1979), based on the Sokoloff model (Sokoloff et al., 1977).

In our PET laboratory, XCT and PET scanners were installed in the same room, so that the patient can be tranferred from XCT to PET gantry with the same head position to ensure comparable slice levels. This enables us to easily determine the region of interest. Seven 16 mm-thick slices were taken at a time. The spacial resolution of this apparatus is 8 mm.

In one case of rigid-type Parkinsonism, a PET scan was performed before and after L-Dopa administration. L-Dopa induced a marked improvement in his behavior.

Cerebral depth recordings were made during the course of stereotactic thalamotomy as described previously (Ohye, 1988). Before the operation, XCT and MRI were taken along the expected tracking plane for thalamotomy. The images thus made give us an idea of the depth structures through which the electrode may pass intraoperatively. In all cases examined in this study the passage of the electrode through the internal segment of pallidum was predicted by an imaging study. Usually, the electrode passes through cortical gray-white matter-Cd head-thalamus, in this sequence. However, in the present cases, the internal segment of Pall was penetrated between Cd and thalamus.

Results

In the normal case, 18FDG accumulation was widely distributed throughout the whole brain (Fig. 1). Quantitative measurement revealed that accumulation of 18FDG increased in the thalamus, basal ganglia and frontal cortex in this order (Fig. 2). In rigid-type Parkinson's disease, in contrast, accumulation of 18FDG was low in general, compared to the normal case. rCBF and rCMRO2 were also low. Especially in the frontal cortical level, 18FDG accumulation was remarkably reduced bilaterally to as low as 20% of the normal case. However, 18FDG in the region of BG was maintained, being almost in normal level or even slightly more. Although comparison between Cd and lenticular nucleus (Put and Pall) was not precise, we have the impression that accumulation in Cd was less than that in the lenticular region in rigid-type PA.

Depth microrecording during the course of stereotactic thalamotomy showed relatively reduced spontaneous activity in Cd, but activity in Pall was substantially high in rigid-type PA (Fig. 3).

In one case of rigid-type PA, $rCMRO_2$ and rCMRGl were compared (examinations were performed at different times) and a reciprocal relation was revealed: $rCMRO_2$ was low, but rCMRGl was high. This relation was also found in other cases of rigid type PA. When the molar ratio (amount of O_2 required to metabolize one mol of glucose) is calculated in such a case, it was 3–4 (normal ratio is 6). This was marked in the striatum of rigid-type PA,

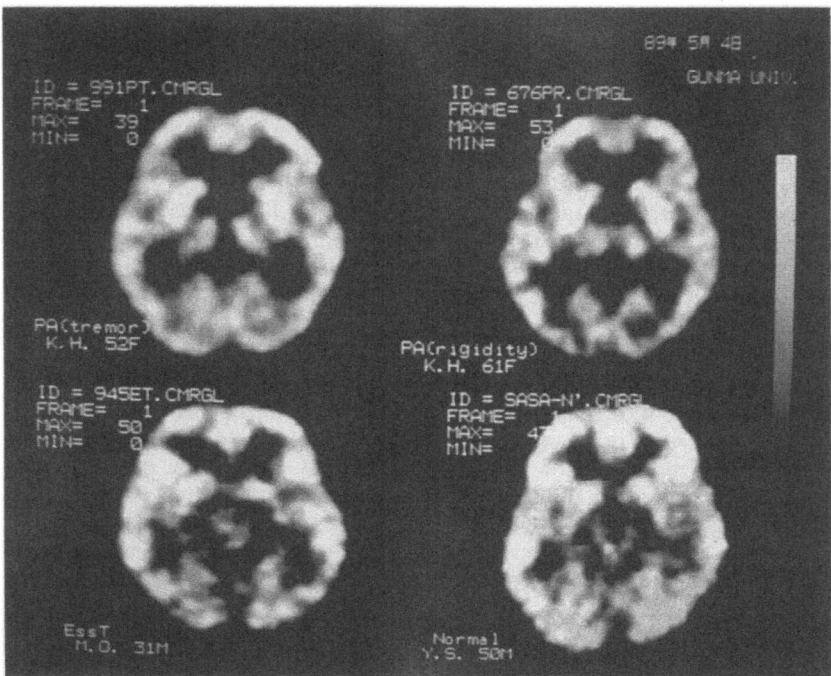

Fig. 1. PET images by FDG to show regional glucose metabolic rate in rigid type Parkinson's disease (upper left), tremor type Parkinson's disease (upper right) essential tremor (lower left) and normal control (lower right). Almost the same slice level containing the basal ganglia is selected and demonstrated together. Note that in Parkinson's disease, cortical rCMRGl is reduced especially in rigid type (original images are in color)

suggesting that anerobic glycolysis might take place instead of the usual aerobic glycolysis.

In another case of a rigid-type PA (juvenile type PA), glucose metabolism was examined before and after administration of L-Dopa. In this case, the Dopa-insufficient state was characterized by strong rigidity (more in the right side) and severe difficulty of movement with poor facial expression. In contrast, L-Dopa administration induced, at first, for about 30 min, severe dyskinesia of choreic movement in his right extremities and then brought about an almost normal state of movement and behavior. In this case, rCMRGl before administration of L-Dopa showed typical PA pattern, being low in the cerebral cortex but relatively high in the left BG. After administration of L-Dopa, with the amelioration of ridigity, and hence his behavior, cortical rCMRGl increased considerably, and rCMRGl in left BG decreased, becoming closer to normal pattern (Fig. 4). At this stage, however, the cortical metabolic rate remained at a level lower than normal. Microrecording during the course of stereotactic

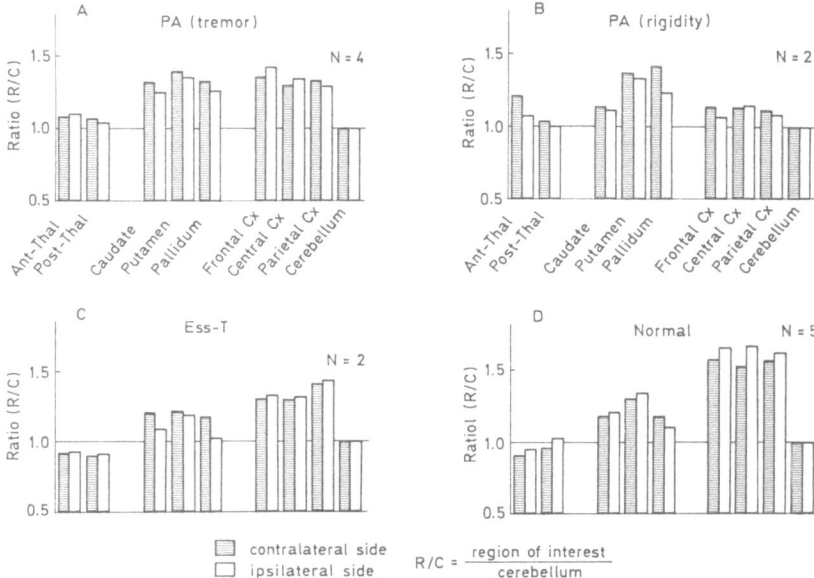

Fig. 2. For four different groups shown in Fig. 1, averaged rCMRGl in different regions of interest is shown by bar graph. Cerebellar metabolic rate is taken to normalize each values. Black bar denotes value in the affected side (contralateral). A Tremor type Parkinson's disease, B Rigid type Parkinson's disease, C Essential tremor, D Normal. In each groups, the left block is rCMRGl of thalamus, middle block is rCHRGl of basal ganglian, right block is rCMRGl of cortex and cerebellum

thalamotomy revealed that Cd activity was relatively low, whereas Pall activity remained rather high.

In tremor-type PA, it was also the case that rCMRGl in BG was high, but rCMRGl in cortical gray was relatively low (see Figs. 1 and 2). However, this change was not as marked as in rigid-type PA. Especially in cortical gray, the decrease in metabolic rate was not obvious. Depth recording in these tremor-type PA showed high activity in Cd, and Pall activity was reduced. No particular discharge pattern such as burst discharges was recorded. Thus in tremor-type as well as in rigid-type PA, rCMRGl and neural activity revealed by depth recording seemed to be parallel.

In a hemi-tremor-type PA (tremor was manifested only in one side) rCMRGl in BG contralateral to the affected limb was elevated, as previously reported by others.

In cases with essential tremor, rCMRGl showed an almost normal pattern (Fig. 1): in the following structure, it increased in this order; thalamus, BG and cerebral cortex (Fig. 2). Depth recording revealed that Cd activity was higher than BG activity, being similar to the result in tremor-type PA.

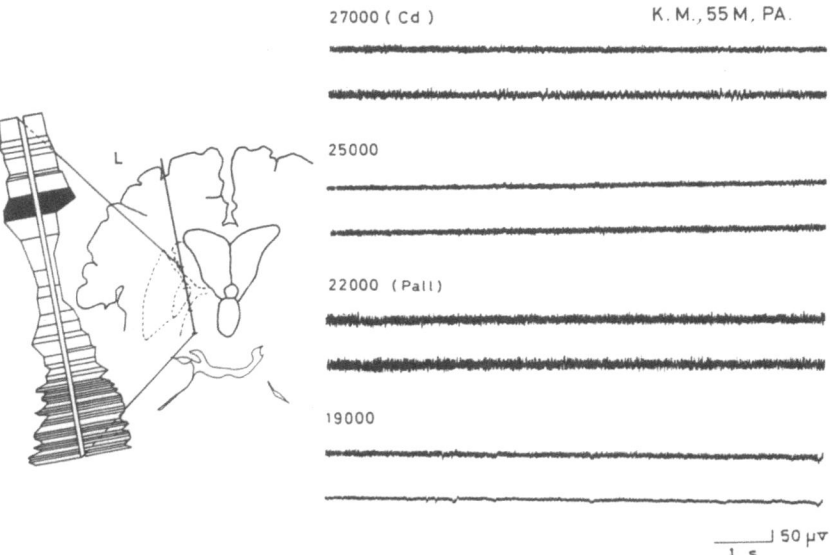

Fig. 3. Electrical activity of internal segment of pallidum (third pair of traces, marked by 22,000), taken from a series of sequential changes of basic background activity along a tracking (shown on the left figure) recorded during a course of stereotactic thalamotomy in a case of pure rigid type Parkinson's disease. Several examples of depth recording are shown on the right side. Numbers in each pair of traces denote distance in microns from zero (end of trajectory drawn on the left). On the left side is shown a brain slice on which recording electrode passes (made from XCT image), and integrated value of electrical activity of basal ganglia along the tracking (enlarged in the extreme left). For depth recording, a pair of electrodes are used in parallel to sagittal plane. Upper trace on the right and left bar graph on the left, from the anteriorly set electrode

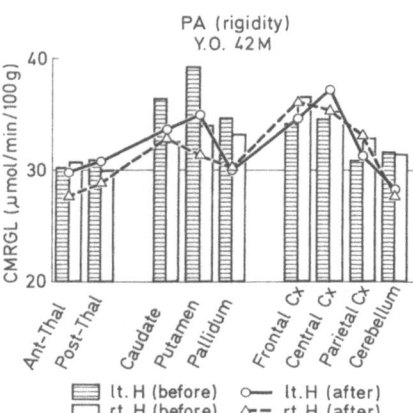

Fig. 4. rCMRG1 in different regions of interest in a case of Parkinson's disease with severe rigidity before (bar graph) and after (line graph) administaration of L-Dopa. Note that rCMRG1 decreased markedly in the region of basal ganglia

Discussion

The present study on the cerebral metabolic rate (oxygen and glucose) and cerebral blood flow using PET scan in Parkinson's desease revealed that the cerebral metabolic rate differ somewhat between tremor-type PA and rigid-type PA. In general, rCMRO$_2$, rCMRG1 and rCBF of the cerebral cortex (especially in frontal and parietal regions) were decreased in PA. Also, the decrease was more marked in rigid-type PA than in tremor-type PA. That is to say that decreased metabolic rate and blood flow in cerebral cortex is a characteristic feature of rigid-type PA. This may be correlated with the clinical feature that in rigid-type PA, the mental as well as physical activity is less vivid, and is not infrequently in a somewhat apallic or inactivated state. In this context, it was interesting to note that in one case of rigid-type PA, the patient's rigid state during Dopa deficiency was correlated with decreased rCMRG1, and L-Dopa ameliorated his physical as well as mental condition together with increase of cortical rCMRG1.

In rigid-type PA, another characteristic feature in the cerebral metabolic rate revealed by this study was that although the cortical rCMRG1 was low, it was maintained high (or normal) in BG, but rCMRO$_2$ decreased in BG. The discrepancy between rCMRG1 and rCMRO$_2$ in BG suggest anerobic glycolysis in BG of rigid-type PA, and should be verified further.

In contrast to rigid-type PA, the cortical metabolic rate and blood flow were not so reduced in tremor-type PA. In this respect, the results of tremor-type PA and essential tremor case were similar or comparable. Therefore, if compared with the normal case, cortical rCMRO$_2$, rCMRG1 and rCBF in tremor-type PA and essential tremor are between normal and rigid-type PA. In other words, from the viewpoint of cortical metabolism and blood flow, tremor-type PA is closer to the normal case than rigid-type PA and comparable to essential tremor.

In the previous reports, Parkinsonian cases were studied as a whole and tremor- and rigid-types were treated together. In our study, we considered these two types separately. Moreover, as rigid-type Parkinsonism, we have selected only severe cases in which the extremities often could not be used properly because of marked rigidity, and certainly L-Dopa therapy was ineffective. These may be the main reason why we were able to detect changes in the metabolic rate not only in unilateral cases but also in bilateral cases, and not only in cortical level but also in BG, where PET study would often fail to reveal any changes.

References

Brooks DJ, Frackowiak RSJ (1989) PET and movement disorders. J Neurol Neurosurg Psychiatry [Special Suppl]:68−77

Eidelberg D, Moeller JR, Dhawan V, Sidtis JJ, Ginos JZ, Strother SC, Cedarbaum J, Greene P, Fahn S, Rottenberg DA (1990) The metabolic anatomy of Parkinson's disease: complementary [18F]fluorodeoxyglucose and [18F]fluoro-dopa positron emission tomographic studies. Movement Disorders 5:203−213

Hirato M, Kawashima Y, Shibazaki T, Shibasaki T, Ohye C (1989) Pathophysiology of parkinsonian rigidity (japanese). Functional Neurosurg 28:1−9

Kuhl DE, Metter EJ, Reige WH (1984) Patterns of local cerebral glucose utilization determined in Parkinson's disease by the [18F]Fluorodeoxyglucose method. Ann Neurol 15:419−424

Leenders KL, Wolfson L, Gibbs JM, Wise RJS, Causon R, Jones T, Legg, NJ (1985) The effects of L-dopa on regional cerbral blood flow and oxygen metabolism in patients with parkinson's disease. Brain 108:171−191

Martin WR (1985) Positron emission tomography in movement disorders. Can J Neurol Sci 12:6−10

Ohye C (1985) Studies on the functional disorders of the brain by PET scan (japanese). Proc 17th Japan Conference on Radiation and Radioisotopes, pp 127−132

Ohye C (1988) Selective thalamotomy for movement disorders: microrecording stimulation, techniques and results. In: Lunsford LD (ed) Modern stereotactic surgery. Martius Nijhoff, Boston, pp 315−331

Ohye C (1989) A new aspect of parkinsonian tremor and rigidity (japanese). Internal Med 63:831−836

Ohye C (1990) Regional cerebral metabolism in movement disorders (japanese). In: Nagatsu T, et al (eds) Receptors of the brain and movement. Heibonsha, Tokyo, pp 135−145

Ohye C, Shibazaki T, Hirato M, Kawashima Y, Matsumura M, Shibasaki T (1990) Neural activity of the basal ganglia in Parkinson's disease studied by depth recording and PET scan. In: Bernardi (ed) Basal Ganglia III. Plenum, New York London, pp 637−644

Phelps ME, Huang SC, Hoffman EJ, Selin C, Sokoloff L, Kuhl DE (1979) Topographic measurement of local cerebral glucose metabolic rate in humans with [F-18]2-fluoro-2-deoxy-D-glucose: validation of method. Ann Neurol 6:371−388

Phelps ME, Maziotta JC (1985) Positron emission tomography: human brain function and biochemistry. Science 228:799−809

Raichle ME, Perlmutter JS, Fox PT (1984) Parkinson's disease: metabolic and pharmacological approaches with positron emission tomography. Ann Neurol [Suppl 15]:S131−132

Rougemeont D, Baron JC, Collard P, Bustany P, Comar D, Agid Y (1984) Local cerebral glucose utilisation in treated and untreated patients with Parkinson's disease. J Neurol Neurosurg Psychiatry 47:824−830

Shibasaki T (1984) Analysis of human cerebral functions using positron emission tomography (PET) − A review (japanese). Progr Neurosci 28:211−228

Shibasaki T (1989) Cerebral functions evaluated by blood flow studies in normal and pathological states (japanese). Neurosci Rev 3:150−178

Sokoloff L, Reivich M, Kennedy C, Des Rosiers MH, Patlak CS, Pettigrew KD, Sakurada O, Shinohara M (1977) The [14C]desoxyglucose method for the measurement of local cerebral glucose utilization: theory, procedure, and normal values in the conscious and anesthetized albino rat. J Neurochem 28:897−916

Wagner HN (1986) Images of the brain: past and prologue. J Nucl Med 27:1929−1937

Wolfson LI, Leenders KL, Brown LL, Jones T (1985) Alterations of regional cerebral blood flow and oxygen metabolism in Parkinson's disease. Neurology 35:1399−1405

Correspondence: Dr. C. Ohye, Department of Neurosurgery, Gunma University School of Medicine, 3-39 Showa-machi, Maebashi, Gunma, Japan.

Role of stereotaxic surgery in treatment of Parkinson's disease

H. Narabayashi

Prof. Emeritus, Juntendo University, and Neurological Clinic, Tokyo, Japan

Summary

Rigidity and tremor are almost completely alleviated by stereotaxic surgery on the ventral subnuclei of the thalamus without producing side effect. With use of micro-electrode technique to analyze the neuronal activity, the VL, which receives the pallidal afferents, is explained responsible for rigidity and the Vim, which receives the proprioceptive afferents from periphery, for tremor. Primary akinesia is not changed at all by surgery but responds well to levodopa. Long-term observations of the postoperative course established that improvement by surgery with/without medicine was more sustained and longlasting in the cases of younger onset than in those of classical PD. This may suggest that in clinical and pathophysiological analysis, the degenerative process of the disease is limited within the nigrostriatal DA system in the former and more widespread in the latter. Within the nigrostriatum, nigroputaminal part is analysed to be first affected. Speed of progression seems to show the difference between cases of early onset parkinsonism and PD.

Introduction

As a method of treatment of Parkinson's disease (PD), stereotaxic surgery is generally considered as a treatment in the past. There is no doubt that the pharmacological approach, especially that with levodopa and related substances, is always the first choice to be tried. Possible application of surgical treatment, however, is usually considered when the pharmacological treatment fails because of uncontrollable reverse effects or poor efficacy. Also there are cases, in which the patient cannot afford to take any antiparkinsonian drug, mostly because of gastrointestinal troubles, the most serious one being ulcer of the stomach or duodenum. In addition, side effects of CNS origin such as psychic symptoms or levodopa-induced dyskinesia preclude continuous use of medicine.

Almost complete relief of parkinsonian rigidity and tremor by stereotaxic surgery has long been known, for the technique was established in the early 1950s (Schaltenbrand and Walker, 1982).

The aim of the stereotaxic technique is to produce highly selective and discrete lesions in certain neural structures in the depth of the brain without causing damage to neighboring structures. The target site may be either a fiber system or a nerve cell group that plays a key role in the production of symptoms such as rigidity and tremor. Throughout the approximately forty years' history of stereotaxic surgery, intensive efforts have been continuously made to improve the technique to acquire the highest degree of accuracy possible in targetting the desired structure.

I. Refinement of the surgical technique

The stereotaxic technique in surgical intervention on the thalamic subnuclei was firstly based on the three dimensional measurement of the target structure in the brain atlas. Radiological measurement referring to visualized figures of the third ventricle, such as the anterior and posterior commissures, gives coordinates of the diencephalic structures and also of the basal ganglia on the brain atlas, mostly the atlas made by Schaltenbrand-Bailey (1959), which was prepared from study of the brain of a healthy young adult. Obviously, such measurements are limited when the patient's brain presents atrophy or dilatation of ventricles due to aging and long-standing disease.

To compensate and overcome such anatomical variability, physiological devices have been developed and used. High-frequency electrical stimulation of 40 to 60 Hz through the electrode at the tip of the insertion needle and reversible cooling of the needle-tip to see changes in symptoms, as well as analysis of the cortical evoked potentials induced by low-frequency stimulation of the various subnuclei of the thalamus (Yoshida et al., 1964), have been introduced. However, the most precise and reliable information to identify the specific structure in the depth of the brain is provided by analysis of the unitary activity of neurons and fibers recorded through a 5–7-micron semimicro-electrode at the tip of the insertion needle as pioneered by Albe-Fessard et al. (1963) in 1963. The detailed description of semimicroelectrode recording technique to register the field potential of neuron activity can be referred to elsewhere (Ohye et al., 1976/77; Ohye, 1982).

When the semimicroelectrode recording technique is used, the following two specially useful and important phenomena are observed. Firstly, the basic noise level, which is the sum of the integrated neuronal activity around the tip-electrode, is delicately different at each depth structure and specially in the subnuclei of the thalamus because of the difference in neuron population density and difference in neuron size (Fukamachi et al., 1973, 1977). As shown in Fig. 1, the highest noise level is obtained when the Vim (ventral intermediate nucleus) is reached in the course of electrode insertion from the frontal burr hole in a slightly medio-ventro-posterior direction. The VL (ventrolateral nucleus) usually presents a lower noise level than the Vim but one higher than that obtained in the dorsal thalamic nuclei.

The second important phenomenon useful for differentiating each sub-nucleus is the observation and analysis of behavior of neurons (Ohye et al.,

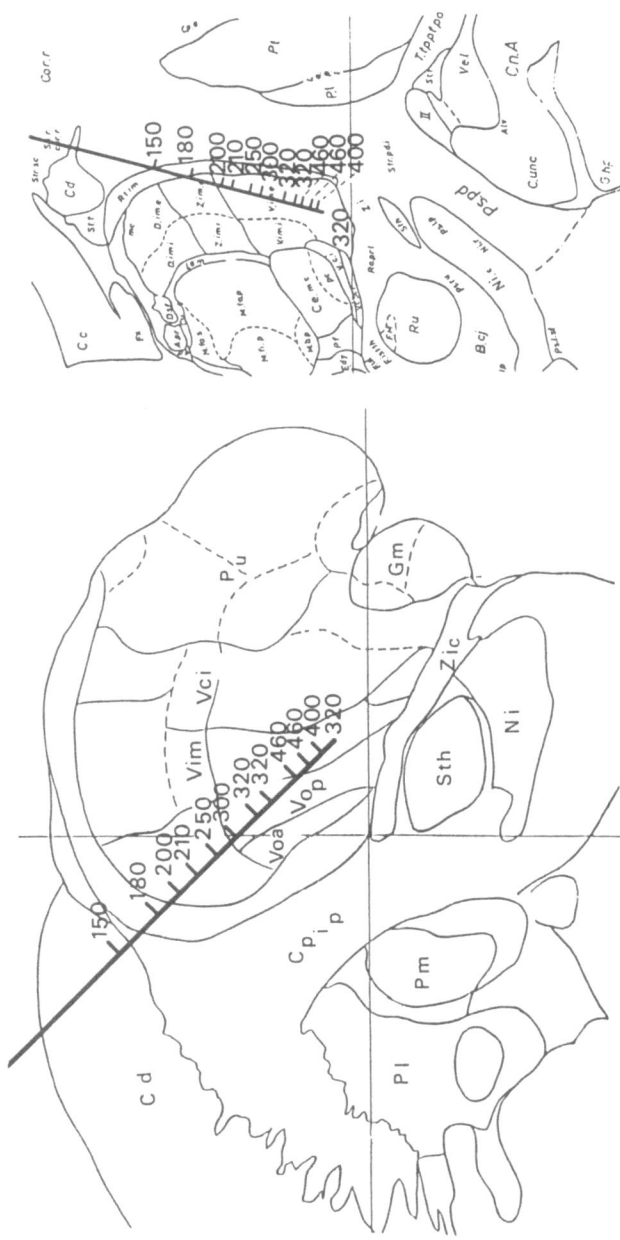

Fig. 1. Parkinsonism. Changes in basic noise level of neuronal activities during the course of semimicroelectrode insertion into the Vim nucleus, projected onto figures from Schaltenbrand and Bailey's brain atlas. (**a**) lateral view (projected on lateral plane 15 mm from midline); (**b**) anteroposterior view (projected on FP7 plane); numbers along needle track indicate level of integrated noise for each 2 ms. From Narabayashi (1990a), reproduced by courtesy of the publisher

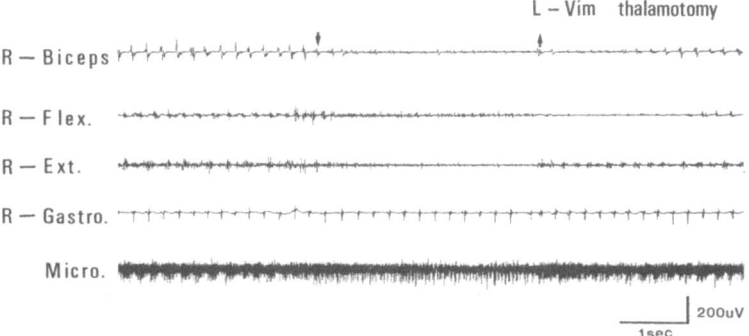

Fig. 2. Tremor-synchronous rhythmic burst discharges in Vim, indicated by Micro. ↑ ↓ :period of passive supression of tremor. *R-Biceps* right biceps brachii muscle; *R-Flex* right forearm flexor muscle; *R-Ext.* right forearm extensor muscle; *R-Gastro.* right gastrocnemius muscle

1974). In the ventral part of the Vim, regular rhythmic bursts of discharges are routinely recordable when the patient presents tremulous movements in the extremities, neck or jaw. These rhythmic discharges are synchronous with tremulous movement in certain parts of the contralateral side of the body, and are interpreted as being evoked by peripheral tremulous movement (Fig. 2). Subthreshold electrical stimulation of the median nerve at the contralateral wrist evokes activity in this neuron area with an eleven to twelve milliseconds latency, which gives the value of conduction velocity of a little more than 70 meters per second, which is as fast as that for proprioceptive impulses.

High-frequency (60Hz) stimulation of these rhythmically discharging or proprioceptive neurons within the Vim always inhibits tremor at the corresponding part of the contralateral extremities. Furthermore, a 3–4 mm diameter lesion produced by electro-thermo-coagulation at 70°C for 30 seconds routinely abolishes severe tremor within seconds on the operating table (Narabayashi and Ohye, 1978; Narabayashi, 1986).

Rigidity can similarly be abolished immediately after placement of a lesion in the ventral part of the VL, which is 2–3 mm anterior to the Vim lesion. The VL receives an afferent inflow from the internal segment of the pallidum, but the Vim does not. These two structures, the ventral parts of the VL and Vim, are located adjacent to each other, but differ in their role in relation to rigidity or tremor (Narabayashi, 1988). However, a lesion in the most ventral part of the Vim, i.e., around the ventral border of the thalamus to the subthalamic field of Forel, can disrupt two inflowing projections, one from the pallidum and the other from the periphery, for these two projections cross one another in this area. Figure 3 indicates such a lesion produced within the Vim in a female case of parkinsonism. Tremor and rigidity were sustainedly alleviated for three and a half years until her death due to pneumonia.

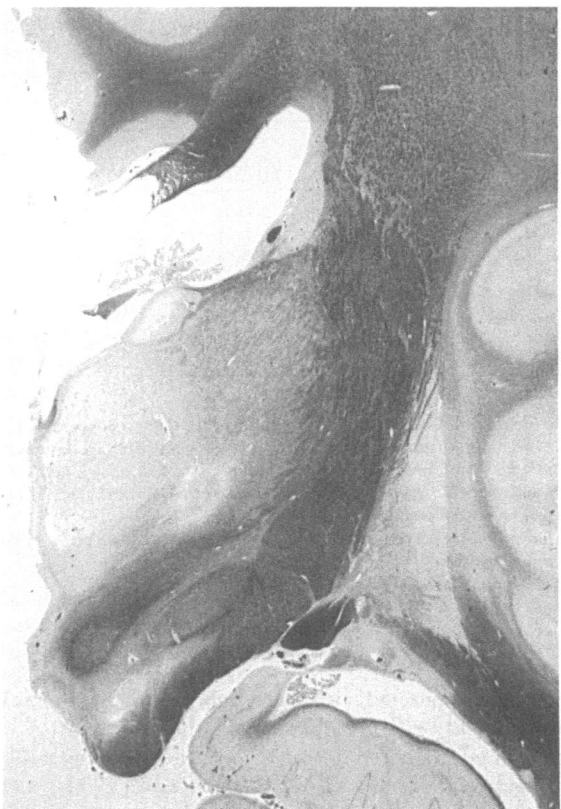

Fig. 3. Frontal section of brain of a female parkinsonian patient who died of pneumonia at age 52, 3.5 years after right-sided Vim & VL thalamotomy. Surgery improved movements of arm and fingers better than leg. Coagulation lesion was produced around neurons responding to wrist movement. From Narabayashi (1988), reproduced by courtesy of the publisher

II. Long-term results of surgical treatment of rigidity and tremor

Improvement of rigidity and tremor is sustained for many years without much recurrence, when the targetted neurons responsible for these two symptoms are precisely targetted and lesioned by use of the microelectrode recording technique. This method has been routinely employed in the author's surgical theatre in all cases without exception since 1972.

There has been no mortality due to surgery in 212 cases treated in the author's clinic from 1977 to 1982 and also in the period after that till today. In addition, no case of sustained capsular paresis or disturbance of speech function

Table 1. Long-term results in tremor-dominant patients operated on unilaterally

Grade	Year of surgery (no. of cases)						Total (21)
	1977 (0)	1978 (2)	1979 (6)	1980 (3)	1981 (8)	1982 (2)	
A	0	2 − (one died of heart attack)	3 −	3 −	7 − (one died of heart attack)	1 −	16 −
B	0	0	2 −	0	0	1 −	3 −
C	0	0	1 −	0	0	0	1 −
D	0	0	0	0	1 +	0	1 +

A: Better ADL than preoperatively; tremor completely stopped on operated side, normal living conditions at jobs or in active housekeeping; progression of tremor on unoperated side nil or slight.
B: Complete alleviation of tremor on operated side; improvement in ADL poorer than in A and similar to preoperative period because of progression of tremor on unoperated side.
C: No tremor on operated side, but disability increased and ADL worse than preoperatively due to progression of akinesia and of tremor on other side.
D: Tremor not controlled by surgery; akinesia generally worse; deterioration in ADL.
−: Tremor remains eliminated.
+: Tremor could not be controlled even at time of surgery.
From Narabayashi (1988)

due to surgery was encountered, although slight, transient, muscular hypotonia for about a week could be found in about 4% of the cases.

The long-term effects noted during a 3−8 years' postoperative observation period, as studied in November 1984 (Narabayashi, 1990a) were previously reported and the details may be found in that report. Briefly, the most

Table 2A. Long-term results in patients with hemiparkinsonism

Grade	Year of surgery (no. of cases)						Total (37)
	1977 (5)	1978 (10)	1979 (4)	1980 (7)	1981 (6)	1982 (5)	
A	2	5	4	6	5	4	26
B	1	3	0	1	0	1	6
C	2	2	0	0	0	0	4
D	0	0	0	0	1	0	1

A: Better ADL than preoperatively; no rigidity nor tremor on operated side, no or slight symptoms on other side.
B: Almost same ADL as preoperatively at time of survey. Operated side, only slight symptoms; symptoms on other side developed.
C: Worse ADL than preoperatively; symptoms on other side and akinesia increased.
D: Not benefited at time of surgery; much deteriorated at time of survey

Table 2B. Hemiparkinsonian patients medicine-free postoperatively and at survey

Case	Year of surgery	(Postop. period, yrs)	Age at onset (yrs)	Duration of illness (yrs)
1	1979	(6)	37	6
2	1979	(6)	57	4
3	1980	(5)	61	5
4	1980	(5)	59	4
5	1981	(4)	43	4
6	1981	(4)	50	5
7	1982	(3)	45	4

From Narabayashi (1988)

important observations are that the postoperative clinical course and pictures seem to differ depending on the age of the patient, the age of the onset of the disease and the years of disease duration, even after similarly marked alleviation of rigidity and tremor.

Table 1 indicates the observation in a series of *tremor dominant cases*. Severe tremor and relative rigidity were sustainedly abolished on the operated side during the observation period with almost no recurrence and the patient remained in much better status than preoperative unless the symptoms on the other side became severer. Sixteen cases (76% of a total of 21 cases) have retained in much better status than the preoperative under the same or lower dose of medicine.

Table 3. Long-term results in patients with asymmetric symptoms

Grade	Year of surgery (no. of cases)						Total (68)
	1977 (3)	1978 (9)	1979 (10)	1980 (10)	1981 (18)	1982 (18)	
A	1	4	5	6	8	15	39
B	1	1	3	3	4	2	14
C	1	2	1	1	6	1	12
D		2	1				3

A: Markedly benefited by surgery; ADL remains improved over preoperative status (independent, working in jobs or active in housekeeping). Disability score at survey, grades I and II or early III.
B: Well benefited by surgery in immediate postoperative period; slowly worse afterward. At time of survey, disability score mostly grade III or early IV, similar to preoperative grade under almost same dose of medicine.
C: Modified relatively by surgery, but at time of survey worse than preoperatively and deteriorated, mostly due to progression of disease. Disability score at time of survey mostly IV.
D: Did not benefit from or not influenced by surgery acutely or subacutely. Due to progression of disease, disability score mostly IV or V.
From Narabayashi (1988)

Table 4. Results in bilaterally operated-on patients

	A (drug-free)	B	Total
Juvenile parkinsonism	8 (2)	3	11

A Almost normal in job and social life.
B Independent at home, full home work

Table 2A presents the similar observations in *cases with hemiparkinsonism*. Twenty-six of the 37 cases remained in a highly improved state at 3−8 years after surgery. Seven of them were almost normal in terms of job and home activities and were still drug free at the time of their last examination (Table 2B).

Table 3 indicates the result of unilateral surgery in the *cases with bilateral symptoms* but with marked asymmetry between the two sides of the body. The cases in this category were the most numerous. The difficulties seen in these cases after the unilateral surgery on the severer side were the existence of primary akinesia, which was not secondary to rigidity and tremor, and the still remaining symptoms on the unoperated side. These patients still required levodopa treatment but the dose of medicine was often lowered to about 20−30% less than the preoperative amount. However, with progression of the disease process over several years, the dose of medicine again needed to be increased to control the advanced akinesia and rigidity of the unoperated side. Even in such an again worsened state, the operated side still presented only minimal symptoms and better functioning, often displaying the reverse asymmetry between the two sides of the body in contrast to the preoperative state.

Cases that initially respond to levodopa markedly and dramatically but later present severe dyskinesia and up-down oscillation of the effects should also be considered for possible surgical treatment, when combination of DA agonist and levodopa proved unsatisfactory to control these reverse effects. As reported by the author and his colleagues, dyskinesia of the extremities is almost completely and sustainedly abolished together with alleviation of rigidity after VL surgery but the dyskinesia in the neck and oro-facial area is usualy not influenced. Use of surgical procedure on such younger onset cases with drug-induced dyskinesia is not well recognized (Narabayashi et al., 1984).

The most important series is the *cases bilaterally operated*, in which rigidity and tremor are both almost completely eliminated on either side of the body. Bilateral intervention should be done very carefully and permitted only exceptionally. The cases should be selected from only the younger onset cases, and strict screening based on careful analysis of data from clinical, neuro-psychological, EEG, and radiological examinations is required. The interval between two-side surgery should be not less than one year.

During the period of 1977–82, only 25 cases were operated-on bilaterally, half for highly incapacitating tremor cases and half for younger onset cases difficult to treat medically, as shown in Table 4. With respect to the group of younger onset cases, all cases presented well and showed almost normal behavior status when examined three to eight years postoperatively. As seen from the Table, 8 cases (A) were all excellent in daily activities and worked normally in jobs and 3 cases (B) stayed at home but well independent of family assistance, mostly under no or a lower dose of medicine than preoperatively (Narabayashi, 1988, 1990a).

III. Lessons from surgical treatment

Precision of lesion placement in the selected structure such as the VL or Vim of the thalamus changed the concept of stereotaxic surgery in practice and made it possible to interpret the clinical effects or changes by the surgery on the neuroanatomical and physiological term.

Lesson 1

The most important finding is the difference between tremor and rigidity. Figure 4 introduces a schematic description of the basal ganglia and its outflow, which underlies the nigrostriatal dopamine system. The efferent projection from the pallidum internum to the VL of the thalamus is experienced and

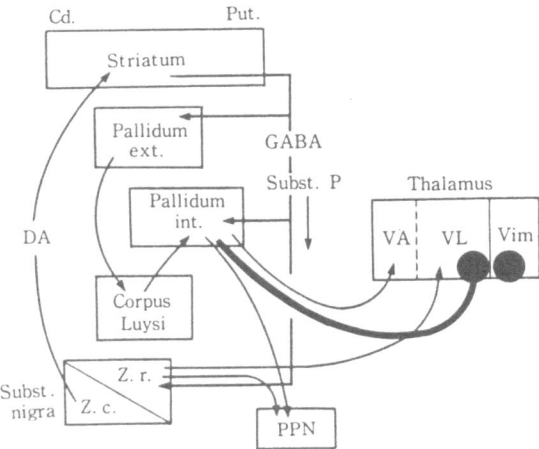

Fig. 4. Schema of projection within the extrapyramidal system producing rigidity (pallidum internum to VL) and tremor (Vim). Black circles indicate the placement of surgical lesions. *Cd* caudate nucleus; *Put* putamen; *PPN* pedunculopontine nucleus

analysed to be responsible for production of rigidity since the first case of pallidotomy by the author in May 1952 (Narabayashi et al., 1956). The Vim lying posterior to the VL and anterior to the VP (ventral posterior nucleus) constitutes the proprioceptive part of the sensory nucleus and is proved and interpreted as being involved in the production of pathological tremor, including both parkinsonian and non-parkinsonian types. Therefore, these two symptoms should be understood as being different in terms of both anatomical basis and physiological generating mechanism.

Lesson 2

Primary akinesia, the third motor symptom after rigidity and tremor, is considered to be different in generation from the above two motor symptoms, since it is not influenced, neither better nor worse, by the Vim and VL surgery that eliminates the other two symptoms. Levodopa, however, works well on both rigidity and primary akinesia and this is the reason the drug therapy predominates over surgical therapy in most cases of PD.

The author has proposed three subtypes of akinesia (Narabayashi, 1980, 1983). Type I is the akinesia secondary to existence of rigidity, which may better be termed as bradykinesia. Type II is the primary akinesia as being discussed and Type III is the akinesia observed in the later, advanced stage of illness. The last one is now being analyzed in relation to NE deficiency, which constitutes another topic in parkinsonism research but is not described further in this paper. Details of these analyses may be referred to in published works by the author and his colleagues (Narabayashi et al., 1981, 1986b; Narabayashi and Kondo, 1987).

Types I and II are both known to depend on the nigrostriatal DA deficiency. Type I akinesia is experienced to be abolished along with elimination of rigidity by surgery, although Type II, primary akinesia, is not changed at all surgically but responds well to levodopa therapy. The reason for the difference has been looked for and one possible explanation is described below.

Lesson 3

Differences in the postoperative course between various groups of patients after similar surgical effects seem to offer a clue for us to understand, at least partly, the reason for the progressive worsening of clinical pictures in the disease.

The group of cases, in which the patients became almost normal and drug free for more than several years after unilateral or bilateral surgical procedure occupies about 6% of all operated-on cases. Akinesia in this group belongs to Type I akinesia or bradykinesia as described. These cases are mostly of those patients relatively younger at the time of surgery and also of younger disease onset (Narabayashi et al., 1986a; Narabayashi, 1990a). In contrast, a group of cases with classical parkinsonism starting after the age of fifty or sixty still need the drug of the same or a little lower dose than preoperatively after the similarly successful results by surgery on rigidity and tremor. Differences between these two groups in postoperative course and in necessity of medication may be related to the difference in underlying pathology.

One case of juvenile onset parkinsonism with dystonia belonging to the former group was pathologically verified after 32 years' course of the disease. The patient started the symptom in the first decade of life and was bilaterally operated-on with an interval of fourteen years between the two operations and became almost normal, requiring a lowered dose of medicine after surgery. In pathology, the nigral neurons were relatively well preserved but showed very few pigmentation. Neuronal loss and glial fiber proliferation in the compact zone of the nigra was seen mostly in its ventrolateral part projecting to the putamen rather than in the part projecting to the caudate, as pointed out by Gibb et al. The details of pathology are described elsewhere (Narabayashi et al., 1986a, 1990b; Gibb et al., 1991).

In the second group of idiopathic parkinsonism, diffuse neuronal degeneration is usually seen in the compact zone, i.e., in both putaminal and caudal parts of the compact zone, although the putaminal part is more severely affected (Goto et al., 1989).

Such a difference in pathology between the two groups may reflect the differences in clinical pictures and in postoperative necessity of levodopa therapy. Difference in the type of akinesia may also be explained in the similar way. Type I akinesia may be due to involvement of the nigroputaminal projection; and Type II to more diffuse nigrostriatal pathology. Progressive worsening of the clinical picture may reflect the gradual spread of pathology from the putaminal part to the entire compact zone, both morphologically and chemically (Narabayashi, 1990b). Surgical procedure is interpreted to work only for symptoms resulting from the nigroputaminal pathology, i.e., rigidity and Type I akinesia with/without tremor. Dopamine compensation therapy may work for all symptoms of nigrostriatal pathology, i.e., rigidity with/without tremor and Type II primary akinesia.

Neurochemical mosaic within the striatum relevant to the efferent and afferent projections is also described (Graybiel and Ragsdale, 1983). A fiber-anatomical study focused on the difference in the efferent pathways from the putamen and from the caudate nucleus are now being intensively investigated.

References

Albe-Fessard D, Arfel G, Guiot G (1963) Activités électriques caractéristiques de quelques structures cérébrales chez l'homme. Ann Chir 17:1185–1214

Fukamachi A, Ohye Ch, Narabayashi H (1973) Delineation of the thalamic nuclei with a microelectrode in stereotaxic surgery for parkinsonism and cerebral palsy. J Neurosurg 39:214–225

Fukamachi A, Ohye Ch, Saito Y, Narabayashi H (1977) Estimation of the neural noise within the human thalamus. Acta Neurochir (Wien) [Suppl 24]:121–136

Gibb WRG, Narabayashi H, Yokochi M, Iizuka R, Lees AJ (1991) New pathological observations in juvenile onset parkinsonism with dystonia (dopa-responsive dystonia). Neurology (in press)

Goto S, Hirano A, Matsumoto S (1989) Subdivisional involvement of nigrostriatal loop in idiopathic Parkinson's disease and striatonigral degeneration. Ann Neurol 26: 766–770

Graybiel AM, Ragsdale CW (1983) Biochemical anatomy of the striatum. In: Emson PC (ed) Chemical neuroanatomy. Raven Press, New York, pp 427–504

Narabayashi H, Okuma T, Shikiba S (1956) Procaine oil blocking of the globus pallidus. Arch Neurol Psychiat 75:36–48

Narabayashi H, Ohye C (1978) Parkinsonian tremor and nucleus ventralis intermedius of the human thalamus. In: Desmedt JE (ed) Physiological tremor, pathological tremors and clonus. Karger, Basel, pp 165–172 (Prog Clin Neurophysiol, vol 5)

Narabayashi H (1980) Clinical analysis of akinesia. J Neural Transm [Suppl 16]:129–136

Narabayashi H, Kondo T, Hayashi A, Suzuki T, Nagatsu T (1981) L-threo-3, 4-dihydroxyphenylserine treatment for akinesia and freezing of parkinsonism. Proc Japan Acad 57 (Ser B):351–354

Narabayashi H (1983) Pharmacological basis of akinesia in Parkinson's disease. J Neural Transm [Suppl 19]:143–151

Narabayashi H, Yokochi F, Nakajima Y (1984) Levodopa-induced dyskinesia and thalamotomy. J Neurol Neurosurg Psychiatry 47:831–839

Narabayashi H (1986) Tremor: its generating mechanism and treatment. In: Vinken PJ, Bruyn GW, Klawans HL (eds) Handbook of clinical neurology, vol 5 (49). Elsevier Science Publishers, Amsterdam, pp 597–607

Narabayashi H, Yokochi M, Iizuka R, Nagatsu T (1986a) Juvenile parkinsonism. In: Vinken PJ, Bruyn GW, Klawans HL (eds) Handbook of clinical neurology, vol 5 (49). Elsevier Science Publishers, Amsterdam, pp 153–165

Narabayashi H, Kondo T, Yokochi F, Nagatsu T (1986b) Clinical effects of L-threo-3, 4-dihydroxyphenylserine in cases of parkinsonism and pure akinesia. In: Yahr MD, Bergmann KJ (eds) Advances in neurology, vol 45. Raven Press, New York, pp 593–602

Narabayashi H, Kondo T (1987) Results of a double-blind study of L-threo-DOPS in parkinsonism. In: Fahn S, Marsden CD, Calne D, Goldstein M (eds) Recent developments in Parkinson's disease, vol 2. Macmillan, New Jersey, pp 279–291

Narabayashi H (1988) Lessons from stereotaxic surgery using microelectrode techniques in understanding parkinsonism. Mt Sinai J Med 55:50–57

Narabayashi H (1990a) Surgical treatment in the levodopa era. In: Stern G (ed) Parkinson's disease. Chapman & Hall, London, pp 597–646

Narabayashi H (1990b) Clinical analysis of juvenile and classical parkinsonism and underlying pathophysiological mechanisms. Presented at the XIth International Congress of Neuropathology, Kyoto, 1990

Ohye Ch, Saito Y, Fukamachi A, Narabayashi H (1974) An analysis of the spontaneous rhythmic and non-rhythmic burst discharges in the human thalamus. J Neurol Sci 22:245–259

Ohye Ch, Maeda T, Narabayashi H (1976/77) Physiologically defined VIM nucleus. Appl Neurophysiol 39:285–295

Ohye C (1982) Depth microelectrode studies. In: Schaltenbrand G, Walker AE (eds) Stereotaxy of the human brain. G Thieme, Stuttgart New York, pp 372–389

Schaltenbrand G, Bailey P (1959) Introduction to stereotaxis with atlas of human brain. G Thieme, Stuttgart New York

Schaltenbrand G, Walker AE (eds) (1982) Stereotaxy of the human brain, 2nd ed. G Thieme, Stuttgart New York

Yoshida M, Yanagisawa N, Shimazu H, Givre A, Narabayashi H (1964) Physiological identification of the thalamic nucleus. Arch Neurol 11:434–443

Correspondence: Prof. H. Narabayashi, Neurological Clinic, 5-12-8 Nakameguro, Meguro-ku, Tokyo 153, Japan.

L-threo-DOPS—implications for pathophysiology of parkinsonian symptoms chronically treated with L-DOPA

T. Kondo

Department of Neurology, Juntendo University School of Medicine, Tokyo, Japan

Summary

During the course of the disease, several L-DOPA-unresponsive symptoms in parkinsonism become manifest. These symptoms, such as frozen gait, disturbance of postural adjustment and bradyphrenia, may be attributed to the abnormality of the nondopaminergic system in the brain.

L-threo-DOPS, a synthesized amino acid, is converted to norepinephrine (NE) by enzymic decarboxylation. Basic and clinical data indicate bifold actions of the substance. The substance permeates into the brain and acts as a precursor of NE not only in peripheral tissues but also in the brain, and enhances release of amines from respective nerve terminals.

The results of clinical investigations indicate that the action of the substance distributes from the reflex mechanism in the spinal cord to mental functions in the brain. Although the underlying mechanism of L-DOPA-resistant symptoms may not be simple, the effect of the substance on frozen gait of Parkinson's disease in advanced stages suggests the importance of the role of the NE system in the disease.

Introduction

Several symptoms of parkinsonian patients in advanced stages, such as frozen gait, disturbance of postural adjustment, and bradyphrenia, are known to be resistant to L-DOPA therapy. It is speculated that these L-DOPA-resistant symptoms may result from nondopaminergic lesions (Bonnet et al., 1987; Narabayashi, 1984; Pillon et al., 1989).

As Narabayashi (1984) has pointed out, akinesia in parkinsonism may be categorized into three types. The first type of akinesia is secondary to rigidity, which is released by stereotactic thalamotomy and of course L-DOPA therapy. The second type of akinesia is poverty of movement or lack of movement, which is improved by L-DOPA therapy, but not by thalamotomy. The third type of akinesia is the freezing phenomenon. This type of akinesia is often observed in cases with advanced parkinsonism under L-DOPA therapy, and in the rare cases with pure akinesia (Imai, 1980; Narabayashi et al., 1976).

From a pharmacological viewpoint, in the former two types of akinesia, dopamine (DA) deficiency in the brain, especially in the striatum, might be responsible for the genesis.

The underlying neuroanatomical and pharmacological genesis of the freezing phenomenon is still poorly understood. Narabayashi et al. (1981) postulated that norepinephrine (NE) deficiency in the brain is responsible for this symptom because the dopamine-β-hydroxylase (DBH) activity in cerebrospinal fluid (CSF) is much lower in the cases with longer duration of illness and the symptom is improved by L-threo-3,4-dihydroxyphenylserine (L-threo-DOPS) administration.

L-threo-DOPS is a synthesized amino acid. After decarboxylation of the substance by aromatic-L-amino-acid decarboxylase (AADC), (-)-NE is formed (Blaschko and Chrusciel, 1960).

In this chapter, the author reviews, first, the clinical and NE correlation in parkinsonism which appears in the literature, then proceeds to the pharmacology of L-threo-DOPS and results of previous clinical investigations.

NE deficiency in the brain and the possible clinical syndrome

Degeneration of neurons in the locus coeruleus (LC) in the parkinsonian brain is very common, as is cell loss in the substantia nigra compacta (Greenfield and Bosanquet, 1953).

Numerous papers have reported a decrement of NE content or lowering of DBH activity (Nagatsu et al., 1984) in the parkinsonian brain. Figure 1 summarizes the distribution of NE and degree of NE deficiency from several papers (Bernheimer et al., 1963; Birkmayer et al., 1975; Ehringer and Hernykiewicz, 1960; Javoy-Agid et al., 1984; Jenner et al., 1983; Kish et al., 1984; Rinne and Sonninen, 1973; Scatton et al., 1983, 1986). The deficiency of NE in the parkinsonian brain is quite extensive, but slight disagreement is seen in some regions.

Javoy-Agid et al. (1984) have reported unchanged hypothalamic NE content and suggested that NE deficiency in the parkinsonian brain does not occur in all parts of the brain where NE bundles are distributed, but only in bundles originating from LC. Cash et al. (1987) also reported that the NE content in LC of nondemented parkinsonian patients was not affected.

Several authors have stressed the importance of NE deficiency in the brain in motor (Barbeau, 1974; Hornykiewicz, 1973; Narabayashi, 1984; Nishi et al., 1987; Pycock et al., 1975) and mental (Cash et al., 1987; Mayeux et al., 1987) impairment in Parkinson's disease. However, only a few papers which correlate clinical symptoms and biochemical evidence can be listed. Mayeux et al. (1987) measured 3-methoxy-4-hydroxyphenylglycol (MHPG), a main metabolite of NE, in CSF and suggested that bradyphrenia in Parkinson's disease may be associated with an alteration in NE metabolism. Stern et al. (1984) suggested

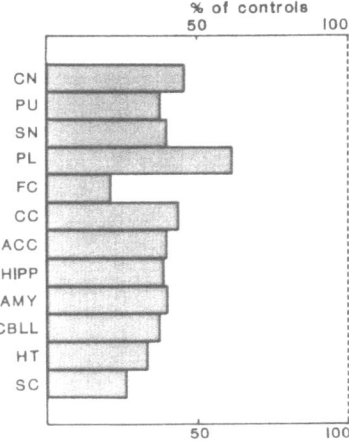

Fig. 1. Degrees of NE decrement in the various brain regions of Parkinson's disease. Values are shown as the average of each percent change against the respective control in the literature. *CN* caudate nucleus, *SN* substantia nigra, *FC* frontal cortex, *ACC* nucleus accumbens, *AMY* amygdalla, *HT* hypothalamus, *PU* putamen, *PL* pallidum, *CC* cingulate cortex, *HIPP* hippocampus, *CBLL* cerebellum, *SC* spinal cord

that the NE system in the brain is related to attention because delayed reaction time correlated with low MHPG level in parkinsonism. Cash et al. (1987) have correlated dementia in living patients and NE content in the LC after the patients' death, and indicated that NE metabolism in the nucleus is affected only in demented parkinsonians.

In relation to motor impairment in parkinsonism, no clinico-biochemical correlative work is noted, except for the previous L-threo-DOPS related studies. Therefore it is very important to elucidate the action of L-threo-DOPS in the brain.

Biochemistry and pharmacology of L-threo-DOPS

L-threo-DOPS is converted into bioactive (-)-NE by the decarboxylation of AADC (Blaschko and Chrusciel, 1960). The substance is generally accepted as a precursor of NE in peripheral tissue (Araki et al., 1981; Reches et al., 1985; Suzuki et al., 1985). Because of its poor permeability into the brain and slow conversion to NE (Bartholini et al., 1975), there was some doubt as to whether L-threo-DOPS plays the role of precursor to NE in CNS (Bartholini et al., 1971; Reches et al., 1985).

Nakamura et al. (1987), using several animal species, demonstrated that L-threo-DOPS penetrates into the brain after intravenous or oral dosing. According to the data, for example in the monkey, L-threo-DOPS distributes almost

Table 1. Concentrations of threo-DOPS and L-DOPA in plasma and CSF. Plasma and CSF were drawn 4 hours after morning administration of L- or DL-threo-DOPS with or without coadministration of L-DOPA during maintenance therapy

Name	Dose of drugs (mg)			Plasma DOPS (µg/ml)	CSF DOPS (µg/ml)	Plasma DOPA (µg/ml)	CSF DOPA (µg/ml)	CSF/Plasma ratio	
	L-, DL-DOPS	DCI	L-DOPA					DOPS	DOPA
T.S.	L-600	B 100	400	1.70	0.03	0.95	0.03	0.018	0.032
T.S.	L-600	C 60	0	1.00	0.09	—	—	0.090	—
M.O.	L-800	B 150	600	1.64	0.07	1.20	0.09	0.043	0.075
K.Y.	L-600	C 30	300	0.85	0.06	0.59	0.02	0.071	0.034
H.I.	L-600	C 60	0	1.53	0.06	—	—	0.039	—
T.S.	DL-1800	C 30	300	2.53	0.28	0.38	—	0.111	—
S.K.	DL-1800	C 15	150	0.73	0.24	0.12	<0.02	0.329	—
A.I.	DL-1800	C 60	0	4.22	<0.02	—	—	—	—

B benserazide, C carbidopa

homogeneously throughout the entire brain, except within the pituitary gland, which has a six to nine times higher concentration of the substance than do other areas. The ratio of the concentration in the monkey brain and that in serum ranged from 0.023 to 0.042 in various brain regions, except for the pituitary gland. The concentration of L-threo-DOPS in CSF of the monkey was similar to that in the brain. A slight species difference was observed in permeability into the brain. The highest uptake of the substance was seen in the cat brain, the next highest in the monkey. Our previous clinical investigation on L-threo-DOPS concentration in CSF and plasma during chronic administration with the substance showed 0.018 to 0.329 as the CSF/plasma ratio (Table 1). This value seems to be similar or much higher than the ratio in the monkey experiment.

The earlier reports indicated that the conversion of L-threo-DOPS to NE was slower than that of L-DOPA (Bartholini et al., 1971). Kato et al. (1987b) demonstrated previously that the substrate affinity (Km) of L-threo-DOPS for AADC is equal to that of L-DOPA and the rate of decarboxylation (Vmax) of L-threo-DOPS is much smaller than that of L-DOPA; the ratio of Vmax of L-threo-DOPS/L-DOPA was about 1/25 in the mouse and 1/100 in the rat brain. The finding on Km corresponds to other data in the same paper; coadministration of L-threo-DOPS with L-DOPA significantly increases the serum concentration of both amino acids, indicating that each amino acid acts as a competitive decarboxylase inhibitor of the other (Kato et al., 1987b).

Naoi and Nagatsu (1986) have shown that monoamine oxidase (MAO) prepared from human placental mitochondria is inhibited by L-threo-DOPS. This may be another reason for the elevated plasma level of L-DOPA and L-threo-DOPS. They have also observed an uptake of L-threo-DOPS into human brain synaptosomes (Naoi and Nagatsu, 1987).

Inconsistent data on the increase in NE content in the brain after administration of L-threo-DOPS have been shown (Bartholini et al., 1971; Kato et al., 1987a; Reches et al., 1985; Toghi et al., 1990). This disagreement, in comparison with the pharmacology of L-DOPA, may arise because of lower penetration through the blood-brain barrier of L-threo-DOPS and a lower Vmax value of L-threo-DOPS for AADC (Kato et al., 1987b). In spite of the poor permeability of L-threo-DOPS into the brain, the apparent content of L-threo-DOPS in the brain may rise due to the lower Vmax, which causes slow conversion from L-threo-DOPS to NE. Hence this slow conversion may be a possible cause of the difficulty in confirming the increase in NE in the brain after administration of L-threo-DOPS. Kato et al. (1987b) observed a significant increase in NE in the brain after administration of 400 mg/kg of L-threo-DOPS with pretreatment of nialamide, an inhibitor of MAO. Brannan et al. (1990) observed an increase in the NE level in the brain after administration of L-threo-DOPS/carbidopa using in vivo dialysis. Another point of disagreement is the change in MHPG. Edwards and Rizk (1981) and Semba and Takahashi (1985) suggested that the rise of the brain MHPG level might be attributed mostly to capillary walls, based on the fact that the rise was diminished by benserazide. However, Kato et al. (1987b) found a dissociation in the suppression of MHPG formation depending on the doses of benserazide between brain and serum. The MHPG formation after administration of L-threo-DOPS in serum was diminished markedly by pretreatment with a low dose (0.3 and 1.0 mg/kg i.p.) of benserazide, but the rise in MHPG in the brain was not modified at these doses. Therefore they concluded that this increase in MHPG was mostly of central origin.

Under various conditions of NE deficiency, supplemental administration of threo-DOPS has been reported with recoveries in the NE content in the brain and improvements of the deficiency syndromes, such as hypothermia by reserpine or tetrabenazine, ptosis by tetrabenazine, and decrease in spontaneous movement by FLA-63 (Kato et al., 1987a; Svensson, 1969), a DBH inhibitor.

Gibson (1988) has found an increase in NE turnover after DL-threo-DOPS administration, and this increase was enhanced in the NE-depleted condition.

Suppressions of harmaline-induced tremor (Yamazaki; et al., 1976) and electroshock seizure (Yoshida and Nakanishi, 1988) by L-threo-DOPS administration have been reported. NE mediated mechanisms are speculated as the causes of these effects.

According to electrophysiological experiments in the reserpinized cat, the action of L-threo-DOPS after conversion to NE in the LC is demonstrated in the striatum (Hirose et al., 1988) or in the spinal trigeminal nucleus (Sasa et al., 1987) when the substance is administered into the ventricles.

L-threo-DOPS, in addition to acting as a precursor of NE in CNS, releases DA, NE and 5-hydroxytryptamine (5-HT) from respective nerve terminals by replacement (Nishino et al., 1987). An inhibition of the release of acetylcholine (Ach) from the nerve terminals by L-threo-DOPS via transsynaptic mechanism after releasing DA or NE from these respective nerve terminals is also known (Nishino et al., 1987).

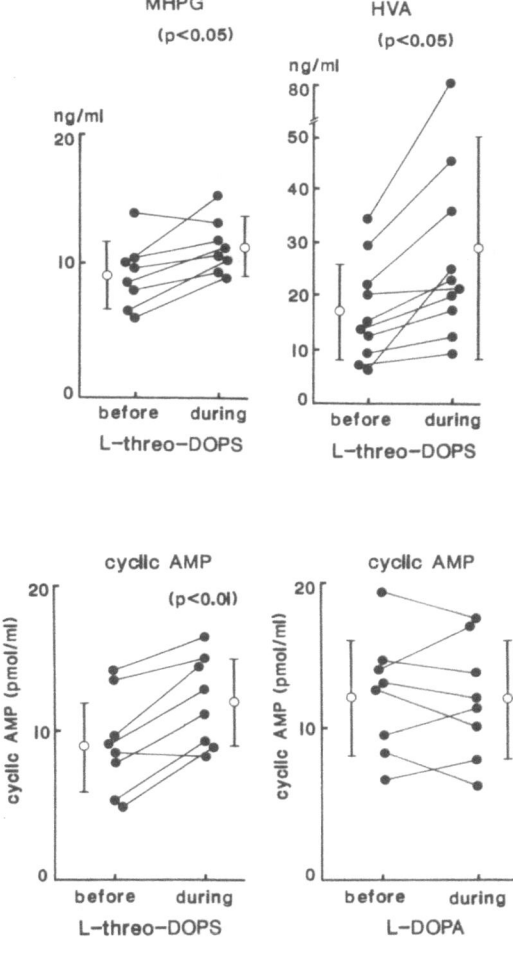

Fig. 2. Changes in CSF cyclic AMP, free MHPG and HVA before and during maintenance therapy of L-threo-DOPS. Dose of L-threo-DOPS ranged from 600 to 900 mg/day with DCI 40 to 60 mg/day. Dose of L-DOPA was 300–500 mg/day with DCI. Paired t-test was employed for statistical analysis

From a clinical viewpoint, we are mostly interested in whether L-threo-DOPS acts as a NE precursor in the brain when the beneficial dose of the substance is administered peripherally for parkinsonian patients.

Several authors have investigated clinicobiochemical effects of threo-DOPS. Suzuki et al. (1984) have found no increase of NE in CSF during repeated administration of the substance. Ogawa et al. (1984) and Yamamoto et al.

(1986) have reported clinical improvement in frozen gait, but the increase of MHPG in CSF has been inconsistent. Teelken et al. (1989) have reported clinical improvements upon administration of DL-threo-DOPS, such as in freezing symptoms, daily activity and mood; however, no increase in MHPG in CSF was observed. Takubo and Kondo (1988) observed the rise in CSF levels of free MHPG and HVA only when L-threo-DOPS/AADC inhibitor(DCI) was administered alone. In the same study, they observed an increase in the cyclic AMP level in CSF during administration of L-threo-DOPS/DCI alone (Fig. 2), or coadministration of L-threo-DOPS and L-DOPA/DCI in the patients medicated with probenecid. As the increase of cyclic AMP in CSF is not influenced by L-DOPA therapy or peripheral cyclic AMP level, the increase indicates some specific action of L-threo-DOPS in CNS.

Recently, Tohgi et al. (1990) have indicated that the concentration of the conjugated form of NE in CSF was 5 to 75 times higher in parkinsonian patients under administration of L-threo-DOPS with or without L-DOPA and DCI. The increase shown under inhibition of peripheral AADC indicates that L-threo-DOPS is converted to NE in CNS.

Table 2 indicates the known actions of L-threo-DOPS. The actions of L-threo-DOPS in CNS may be characterized as twofold: it functions both as a precursor of NE in CNS and as a releaser of monoamines.

Clinical investigations on L-threo-DOPS

The results of a nationwide double-blind trial of L-threo-DOPS for Parkinson's disease and pure akinesia have been reported in Japan (Narabayashi et al., 1987; Narabayashi and Kondo, 1987). Significant improvements seen in the report are frozen gait, especially in the case with kinesie paradoxale, freezing in speech, retropulsion and orthostatic dizziness in Parkinson's disease. According to the analysis of global improvement rating classified by patient background, the patients aged 60 to 69 with severity in Yahr's stage III, clinical course between 5 to 10 years, and age at onset of over 45 were shown to be most responsive to L-threo-DOPS therapy. The effective dose shown by the analysis was 600 mg per day or more. The analysis on the patients with pure akinesia was not significant because the number of patients was so small. Four out of the 14 L-threo-DOPS-treated patients experienced moderate improvement, while 0 out of 10 patients receiving placebos improved.

Several clinicophysiological and neuropsychological analyses on the effects of L-threo-DOPS have been reported. These results might be helpful in understanding the underlying mechanism and sites of action of L-threo-DOPS.

Reaction time has been used as a parameter to measure the impairment of psychomotor function in Parkinson's disease. Yokochi et al. (1986) measured the reaction time in Parkinson's disease and control subjects with no neuropsychological disease, before and after single doses of L-threo-DOPS. The reaction time was delayed in younger control subjects and younger patients of Parkinson's disease by L-threo-DOPS administration, although the reaction

Table 2. Known actions of L-threo-DOPS in the brain. Results in vivo experiments are listed only for the cases in which L-threo-DOPS was administered peripherally, except for in electrophysiological experiments

Investigations	Results	References
Mitochondrial MAO	inhibition	Naoi and Nagatsu (1986)
Blood brain barrier permeability	yes about 1/4 of L-DOPA diffuse distribution in the brain	Nakamura et al. (1987)
Conversion to NE by AADC	yes Km: equal to L-DOPA Vmax: 1/25 ~ 1/100 of L-DOPA	Kato et al. (1987)
Uptake into synaptosome	yes	Naoi and Nagatsu (1987)
Uptake into neuron	LC neuron aminergic neurons	Hirose et al. (1988) Sasa et al. (1987)
Transmitter release	enhancement: NE, DA, 5-HT by displacement inhibition: Ach secondary to release of NE, DA	Nishino et al. (1987)
CSF chemistry (human)	conjugate NE increase MHPG inconsistent among authors cyclic AMP increase HVA increase	Tohgi et al. (1990) Ogawa et al. (1984) Takubo and Kondo (1988) Teelken et al. (1989) Yamamoto et al. (1986) Takubo and Kondo (1988) Takubo and Kondo (1988)
Recovery from NE depletion	free NE inconsistent among authors	Kato et al. (1987a) Reches et al. (1985)
Increase in NE	free NE brain microdialysis	Brannan et al. (1990)
Increase in NE metabolite	MHPG depends on dose of DCI	Edwards and Rizk (1981) Kato et al. (1987b)

time in aged controls was not influenced and that in aged patients with Parkinson's disease was shortened. According to their speculation, in the younger control and younger parkinsonism patients, the basal arousal level was normal, but was elevated excessively after L-threo-DOPS administration, resulting in delayed reaction time. In the aged control, the basal arousal level was subnormal; hence the change of reaction time after L-threo-DOPS administration was inconsistent. In the aged parkinsonian patients, the basal arousal level was low, but was normalized after L-threo-DOPS administration, resulting in a normalized reaction time. On the bases of such age-dependent change in reaction time influenced by L-threo-DOPS, they postulated that an optimal arousal level was needed to initiate movement with shorter latency. These effects of L-threo-DOPS were similar to the raising of the arousal level by extrinic warning stimulation. They stated that the mechanism of effects of L-threo-DOPS on the reaction time was caused by elevation of the arousal level.

The P-300 has been utilized as an electrophysiological index of cognitive function. Yokota et al. (1990) demonstrated that the latency of P-300 is significantly shortened after administration of L-threo-DOPS compared with the latency under L-DOPA therapy or placebo administration in the parkinsonian patients.

From the results of frequency analysis of electroencephalography (EEG) in the normal subjects, the frequency of alpha activity in the occipital region was reduced by acute administration of L-threo-DOPS but unchanged by chronic administration (Nakamura et al., 1987). On the other hand, Komiya et al. (1988) have reported a case study who had hypersomnia and subcortical dementia caused by bilateral medial thalamic and midbrain infarcts treated by L-threo-DOPS. Clinical improvements were observed in hypersomnia, spontaneity and mental tests accompanied by an increase in the fast alpha component in EEG.

Teramoto et al. (1987) have reported the results of psychological tests examined before and during L-threo-DOPS in 8 cases with Parkinson's disease. According to the report, the average of Hamilton's scale for depression in 8 cases improved from 20 to 12. The WAIS score was elevated both in the verbal score and performance score in all patients examined. The pedometer count, which was employed as a parameter of akinesia or total movement in daily life, was increased in 6 out of 8 cases.

Improvements of dementia by L-threo-DOPS administration in Alzheimer-type dementia and dementia accompanied with cerebrovascular disease have also been reported (Ohtsuka et al., 1988).

The dysfunction of the monoaminergic system in the brain has been speculated to be a possible cause of sleep apnea syndrome. Inoue et al. (1989) have observed an effect of L-threo-DOPS/DCI on sleep apnea.

Ueno (1989) has observed an improvement of frozen gait in 5 out of 7 patients with cerebral multiple infarction after administration of 600 mg of L-threo-DOPS for 4 weeks; this improvement was accompanied by an increase in the cerebral blood flow of the frontal lobe in 3 out of 5 patients examined.

The influence of L-threo-DOPS administration on the H-reflex of the soleus muscle has been examined by Nakajima et al. (1987). The depression of the H wave after conditioning group Ia volley, the cause of which is explained as transmitter depletion, was reduced by L-threo-DOPS administration in patients with Parkinson's disease as well as in control subjects.

These clinical investigations on L-threo-DOPS indicate that the effects of the substance are extensive, distributing anatomically from spinal cord to cerebral cortex, or functionally from the reflex mechanism in the spinal cord to motor and psychiatric function in the brain. Moreover, as suggested by the results of reaction time or frequency analysis of EEG, clinical effects of L-threo-DOPS may not be uniform, depending on the conditions of subjects.

Anatomical and pharmacological consideration of frozen gait

The mechanism of the freezing phenomenon is poorly understood. The symptoms appear in patients with Parkinson's disease and in patients with pure akinesia. In the case of Parkinson's disease, the symptoms appear at late stages of the disease under L-DOPA therapy. In the case of pure akinesia, the freezing phenomenon, especially frozen gait, is the main symptom of the disease.

The L-DOPA-resistant symptoms, such as the freezing phenomenon, disturbance of postural adjustment, bradyphrenia and dysarthria seen in late stages of Parkinson's disease, have been postulated to be nondopaminergic dysfunctions (Bonnet et al., 1987; Narabayashi, 1984; Pillon et al., 1989). The deficiency of NE in the brain is well documented in Parkinson's disease as indicated in Fig. 1, although the chemical evidence of NE deficiency in the brain is poorly demonstrated in pure akinesia (Imai, 1980). The contribution of NE dificiency in the brain to the above-mentioned symptoms has been considered.

On the basis of the effects of L-threo-DOPS, Narabayashi (1984) postulated that the NE deficiency in the parkinsonian brain is responsible for freezing symptoms. In a previous report, from clinical observation, the author indicated two kinds of effects of threo-DOPS, i.e., one as a precursor of NE in CNS and the other, facilitative modification on DA turnover, because rigidity and bradykinesia have been improved slightly by administration of threo-DOPS (Kondo, 1984). As indicated by previous clinicopharmacological investigations, L-threo-DOPS permeates into the brain and is converted to NE and enhances DA turnover, resulting in an increase in the conjugated NE (Tohgi et al., 1990), cyclic AMP and HVA (Takubo and Kondo, 1988) in CSF. In the cases of Parkinson's disease in late stages treated by L-DÓPA, further administration of L-DOPA may result in an unfavorable effect on the freezing symptoms. In the case of pure akinesia, L-DOPA therapy is ineffective. Therefore the effect of L-threo-DOPS on those L-DOPA-resistant symptoms should be attributed to the increase in NE in CNS other than that in DA.

Frozen gait and postural instability seen in Parkinson's disease partly resemble those seen in cerebrovascular disorders, such as Binswanger's disease or the lacunar state.

Difficulty in repetitive movement is represented by the freezing or hastening phenomenon in advanced parkinsonism. Nakamura et al. (1978) have analyzed the freezing phenomenon using the finger-tapping test. According to their analysis, the freezing phenomenon is attributed to a disturbance in rhythm formation. The structures responsible for the phenomenon were proposed to be the fronto-striatal complex (Nagasaki et al., 1981).

Yanagisawa (1985) has pointed out the similarity in motor and mental disturbance of the patients with frontal lobe disorder and the patients with Parkinson's disease, including frozen gait, difficulty in initiating and ending motion, slowness of movement, and unconcern. Ueno (1989) has also presented a similar contribution in comparative analysis of the patients with frozen gait due to multiple cerebral infarction or Parkinson's disease, and the patients with gait apraxia caused by mediobasal infarction of the frontal lobe.

Since distribution of NE deficiency in the cerebral cortex is extensive and corticostriatal connection is massive in the striatum (Parent, 1990), it is possible to speculate that NE deficiency in the frontal cortex causes disorganization of the cortico-striatal connection, resulting in both motor and mental disorders in Parkinson's disease. However, as NE deficiency in CNS in Parkinson's disease is widely distributed, other regions, such as the spinal cord and cerebellum, which receive NE bundles from LC, may also be considered to be responsible for generation of L-DOPA-resistant motor disturbances of the disease.

Conclusion

As mentioned above, the role of L-threo-DOPS in the brain has been elucidated. It seems that L-threo-DOPS acts as a precursor of NE in the brain.

According to the results of a double-blind study of L-threo-DOPS, the rate of improvement in frozen gait was about 27 percent of the patients with kinesie paradoxale (Narabayashi et al., 1987). Tohgi et al. (1990) observed an increase in the conjugated NE during L-threo-DOPS therapy, although this increase was not parallel to the clinical improvement. Therefore NE deficiency may not be totally responsible for the generation of freezing symptoms. However, the effects of L-threo-DOPS on Parkinson's disease in late stages suggested the importance of the NE system in the brain in motor and mental functions.

Further clinicochemical or clinicopharmacological investigations should be carried out to elucidate the mechanism of the symptoms seen in advanced Parkinson's disease.

References

Araki H, Fujiwara H, Tanaka C (1981) Age-related changes in the chronotropic effect and the enzymic decarboxylation of L-threo-3,4-dihydroxyphenylserine in the rat heart. J Pharm Pharmacol 33:778–782

Barbeau A (1974) The clinical physiology of side effects in long-term L-DOPA therapy. In: McDowell F H, Barbeau A (eds) Advances in neurology 5. Raven Press, New York, pp 347–365

Bartholini G, Constantidinis J, Puig M, Tissot R, Pletscher A (1975) The stereoisomers of 3,4-dihydroxyphenylserine as precursors of norepinephrine. J Pharmacol Exp Ther 193:523–532

Bartholini G, Constantinidis J, Tissot R, Pletscher A (1971) Formation of mono-amines from various amino acids in the brain after inhibition of extracerebral decarboxylase. Biochem Pharmacol 20:1243–1247

Bernheimer H, Birkmayer W, Hornykiewicz O (1963) Zur Biochemie des Parkinson - Syndroms des Menschen–Einfluß der Monoaminoxydase-Hemmer-Therapie auf die Konzentration des Dopamins, Noradrenalins und 5-Hydroxytryptamins im Gehirn. Klin Wschr 41:465–469

Birkmayer W, Danielzyk W, Neumayer E, et al. (1975) Dopaminergic supersensitivity in parkinsonism. Adv Neurol 9:121–129

Blaschko H, Chrusciel TL (1960) The decarboxylation of amino acids related to tyrosine and their awakening action in reserpine-treated mice. J Physiol 151:272–284

Bonnet A-M, Loria Y, Saint-Hilaire M-H, Lhermitte F, Agid Y (1987) Does long-term aggravation of Parkinson's disease result from nondopaminergic lesions? Neurology 37:1539–1542

Brannan T, Bhardwaj A, Yahr MD (1990) L-threodops increase extracellular norepine-phrine levels in the brain: an in vivo study. Neurology 40:1134–1135

Cash R, Dennis T, L'Heureux R, Raisman R, Javoy-Agid F, Scatton B (1987) Parkinson's disease and dementia: norepinephrine and dopamine in locus ceruleus. Neurology 37:42–46

Edwards DJ, Rizk M (1981) Effects of amino acid precursors on catecholamine synthesis in the brain. Prog Neuropsychopharmacol 5:569–572

Ehringer H, Hornykiewicz O (1960) Verteilung von Noradrenalin und Dopamin (3-Hydroxytyramin) im Gehirn des Menschen und ihr Verhalten bei Erkrankungen des extrapyramidalen Systems. Klin Wschr 38:1236–1239

Gibson CJ (1988) Increase in norepinephrine turnover after tyrosine or DL-threo-3,4-dihydroxyphenylserine (DL-threo-DOPS). Life Sci 42:95–102

Greenfield JG, Bosanquet FD (1953) The brain-stem lesions in parkinsonism. J Neurol Neurosurg Psychiatry 16:213–226

Hirose A, Sasa M, Ohno Y, Takaori S (1988) Inhibitory effects of L-threo-DOPS, an L-noradrenaline precursor, on locus coeruleus-originating neurons in the caudate nucleus. Jpn J Pharmacol 48:435–440

Hornykiewicz O (1973) Dopamine in the basal ganglia – its role and therapeutic implications (including the clinical use of L-DOPA). Br Med Bull 29:172–178

Imai H (1980) Syndrome of pure akinesia or freezing phenomenon without rigidity and tremor and with no effect by L-DOPA therapy. Adv Neurol Sci 24:838–848

Inoue Y, Hoshino E, Sakamoto I, Yamashita Y, Ueda K, Hamazoe K, Fukuma E, Hazama H (1989) The effect of L-threo-3,4-dihydroxyphenylserine (L-threo-DOPS) on sleep apnea syndrome. Jpn J Neuropsychopharmacol 11:841–850

Javoy-Agid F, Ruberg M, Pique L, Bertagna X, Taquet H, Studler JM, Cesselin F, Epilbaum J, Agid Y (1984) Biochemistry of the hypothalamus in Parkinson's disease. Neurology 34:672–675

Jenner P, Sheehy M, Marsden CD (1983) Noradrenaline and 5-hydroxytryptamine modulation of brain dopamine functions for the treatment of Parkinson's disease. Br J Clin Pharmacol 15:277s–289s

Kato T, Katsuyama M, Karai N, Nakamura M, Katsube J (1987a) Studies on the central action of L-threo-3,4-dihydroxyphenylserine (L-threo-DOPS) in FLA-63-treated mice. Pharmacol Biochem Behav 26:407–411

Kato T, Karai N, Katsuyama M, Nakamura M, Katsube J (1987b) Studies on the activity of L-threo-3,4-dihydroxyphenylserine (L-DOPS) as a catecholamine precursor in the brain: comparison with that of L-DOPA. Biochem Pharmacol 36:3051–3057

Kish SJ, Shannak KS, Rajput AH, Gilbert JJ, Hornykiewicz O (1984) Cerebellar norepinephrine in patients with Parkinson's disease and control subjects. Arch Neurol 41:612–614

Komiya T, Narabayashi H, Kondo T (1988) The effectivity of L-threo-3,4-dihydroxyphenylserine (L-threo-DOPS) to the hypersomnia and subcortical dementia caused by bilateral medial thalamic and midbrain infarcts. Clin Neurol 28:268–271

Kondo T (1984) DL-threo-3,4-dihydroxyphenylserine (DL-threo-DOPS) treatment on the patients with Parkinson's disease or pure akinesia. Clin Neurol 24:280–288

Mayeux R, Stern Y, Sano M, Cote L, Williams JBW (1987) Clinical and biochemical correlates of bradyphrenia in Parkinson's disease. Neurology 37:1130–1134

Nagasaki H, Kosaka K, Nakamura R (1981) Disturbance of rhythm formation in patients with hemispheric lesion. Tohoku J Exp Med 135:231–236

Nagatsu T, Yamaguchi T, Rahman MK, Trocewicz J, Oka K, Nagatsu I, Narabayashi H, Kondo T, Iizuka R (1984) Catecholamine-related enzymes and the biopterin cofactor in Parkinson's disease and related extrapyramidal diseases. Adv Neurol 40:467–473

Nakajima Y, Kagamihara Y, Nagaoka M, Tanaka R (1987) The effect of L-threo-DOPS on synaptic transmission to soleus motoneuron in normal subjects and patients with Parkinson's disease. Neurosci Res 5:16–27

Nakamura R, Nagasaki H, Narabayashi H (1978) Arrhythmokinesia in parkinsonism. In: Birkmayer W, Hornykiewicz O (eds) Advances in parkinsonism. Roche, Basel, pp 258–268

Nakamura M, Kato T, Katsuyama M, Karai N, Kumasaka M, Shohno F (1987) Penetration of L-threo-DOPS, a precursor of norepinephrine, into the brain in various experimental animals. Jpn Pharmacol Ther 15 [Suppl 2]:119–125

Nakamura J, Kohgasa H, Hashimoto M, Furatsu M, Mitsune K, Horikawa Y, Inanaga K (1987) The effect of L-threo-DOPS on human CNS. Jpn Pharmacol Ther 15 [Suppl 2]:153–160

Naoi M, Nagatsu T (1986) Inhibition of monoamine oxidase by 3,4-dihydroxyphenylserine. J Neurochem 47:604–607

Naoi M, Nagatsu T (1987) Uptake of L-threo-dihydroxyphenylserine into human brain synaptosomes. J Neural Transm 70:51–61

Narabayashi H, Nakanishi T, Yosida M, Yanagizawa N, Mizuno Y, Kanazawa I, Kondo T (1987) Therapeutic effects of L-DOPS in Parkinson's disease – double-blind, comparative study against placebo as control in patients with the long-term levodopa therapy. Clin Eval 15:423–457

Narabayashi H, Kondo T (1987) Results of a double-blind study of L-threo-DOPS in Parkinsonism. In: Fahn S, Marsden CD, Calne D, Goldstein M (eds) Recent developments in Parkinson's disease. Macmillan Healthcare information, Florham Park New Jersey, pp 279–291

Narabayashi H, Imai H, Yokochi M, Hirayama K, Nakamura R (1976) Cases of pure akinesia without rigidity and tremor and with no effect by L-DOPA therapy. In: Birkmayer W, Hornykiewitz O (eds) Advances in parkinsonism. Roche, Basel, pp 335–342

Narabayashi H, Kondo T, Hayashi A, Suzuki T, Nagatsu T (1981) L-threo-3,4-dihydroxyphenylserine treatment for akinesia and freezing of parkinsonism. Proc Jpn Acad 57 (Ser B):351–354

Narabayashi H (1984) Akinesia in parkinsonism – clinical and pharmacological analysis of parkinsonian symptoms. Bull Acad Med Bel 139:309–320

Nishi K, Kondo T, Narabayashi H (1987) A mouse model of N-methyl-4-phenyl-1,2,3,6-tetrahydropyridine induced parkinsonism: effect of norepinephrine terminal destruction. Brain and Nerve 39:663–672

Nishino K, Fujii Y, Kondo M, Shuntoh H, Fujiwara H, Tanaka C (1987) Effects of L-threo-3,4-dihydroxyphenylserine on efflux of monoamines and acetylcholine in guinea pig brain. J Pharmacol Exp Ther 242:621–628

Ogawa N, Kuroda H, Yamamoto M, Nukina I, Ota Z (1984) Improvement in freezing phenomenon of Parkinson's disease after DL-threo-3,4-dihydroxyphenylserine. Acta Med Okayama 38:301–304

Ohtsuka M, Shimizu N, Dohbutsu M, Saitoh T, Kaneko J, Jyoshita Y, Mizuno Y, Yoshida M (1988) L-threo-3,4-dihydroxyphenylserine treatment for dementia of various etiologies. Neurol Ther 6:355–360

Parent A (1990) Extrinsic connections of the basal ganglia. TINS 13:254–258

Pillon B, Dubois B, Cusimano G, Bonnet A-M, Lhermitte F, Agid Y (1989) Does cognitive impairment in Parkinson's disease result from non-dopaminergic lesions? J Neurol Neurosurg Psychiatry 52:201–206

Pycock CJ, Donaldson I, MacG, Marsden CD (1975) Circling behaviour produced by unilateral lesions in the region of the locus coeruleus in rats. Brain Res 97:317–329

Reches A, Jackson-Lewis V, Fahn S (1985) DL-threo-DOPS as a precursor of noradrenaline. Arch Pharmacol 331:202–208

Rinne UK, Sonninen V (1973) Brain catecholamines and their metabolites in parkinsonian patients. Arch Neurol 28:107–110

Sasa M, Ohno Y, Nabatame H, Yoshimura N, Takaori S (1987) Effects of L-threo-DOPS, an L-noradrenaline precursor, on locus coeruleus-originating neurons in spinal trigeminal nucleus. Brain Res 420:157–161

Scatton B, Javoy-Agid F, Rouquier L, Dubois B, Agid Y (1983) Reduction of cortical dopamine, noradrenaline, serotonin and their metabolites in Parkinson's disease. Brain Res 275:321–328

Scatton B, Dennis T, L'Heureux R, Monfort JC, Duyckaerts C, Javoy-Agid F (1986) Degeneration of noradrenergic and serotonergic but not dopaminergic neurones in the lumber spinal cord of parkinsonian patients. Brain Res 380:181–185

Stern Y, Mayeux R, Cote L (1984) Reaction time and vigilance in Parkinson's disease. Possible role of altered norepinephrine metabolism. Arch Neurol 41:1086–1089

Semba J, Takahashi R (1985) The effects of L-threo-3,4-dihydroxyphenylserine on norepinephrine metabolism in rat brain. Psychiatr Res 15:319–326

Suzuki T, Sakoda S, Ueji M, Kishimoto S (1985) Mass spectrometric measurements of norepinephrine synthesis in man from infusion of isotope-labelled L-threo-3,4-dihydroxyphenylserine. Life Sci 36:435–442

Suzuki T, Sakoda S, Ueji M, Kishimoto S, Hayashi A, Kondo T, Narabayashi H (1984) Treatment of parkinsonism with L-threo-3,4-dihydroxyphenylserine: a pharmacokinetic study. Neurology 34:1446–50

Svenson TH (1969) On the significance of control noradrenalin for motor activity. Experiments with new dopamine-ß-hydroxylase inhibitor. Eur J Pharmacol 7:278–282

Takubo H, Kondo T (1988) Effects of L-threo-DOPS on monoamine metabolites and cyclic AMP in human cerebrospinal fluid – the evaluation of the mechanism for freezing phenomenon and akinesia. Jpn J Neuropsychopharmacol 10:159–168

Teelken AW, Berg GA van den, Muskiet FAJ, Staal-Schreinemachers AL, Wolthers BG, Lakke JPWF (1989) Catecholamine metabolism during additional administration of DL-threo-3,4-dihydroxyphenylserine to patients with Parkinson's disease. J Neural Transm 1:177–188

Teramoto H, Yokochi M, Kujirai T, Aida K (1987) Clinical evaluation of L-thero-DOPS in parkinsonism with special reference to neuropsychiatric study of 8 cases. Jpn J Clin Psychiat 16:765–774

Tohgi H, Abe T, Takahashi S, Takahashi J, Ueno M, Nozaki Y (1990) Effect of a synthetic norepinephrine precursor, L-threo-3,4-dihydroxyphenylserine of the total norepinephrine concentration in the cerebrospinal fluid of parkinsonian patients. Neurosci Lett 166:194–197

Ueno E (1989) Clinical and physiological study of apraxia of gait and frozen gait. Clin Neurol 29:275–283

Yamamoto M, Ogawa N, Ujike H (1986) Effect of L-threo-3,4-dihydroxyphenylserine chronic administration on cerebrospinal fluid and plasma free 3-methoxy-4-hydroxyphenylglycol. Concentration in patients with Parkinson's disease. J Neurol Sci 73:39–44

Yamazaki M, Ikeda Y, Ishikawa M, Inagaki C, Tanaka C (1976) Inhibition of harmaline tremor induced by L-threo-3,4-dihydroxyphenylserine, a norepinephrine precursor. Folia Pharmacol Japon 72:363–369

Yanagisawa N (1985) Disorders of voluntary movements in Parkinson's disease. Jpn J Neuropsychopharmacol 7:817–825

Yokochi F, Nakamura R, Narabayashi H (1986) Basal ganglia and arousal – reaction time study in parkinsonian patients. Adv Neurol Sci 30:841–846

Yokota J, Ito K, Imai H, Narabayashi H (1990) The effect of L-threo-DOPS on P-300 in parkinsonism. Clin Neurol 30:499–504

Yoshida M, Nakanishi T (1988) Inhibitory effects of L-threo-DOPS on electroshock seijure in mice. Brain and Nerve 41:567–573

Correspondence: T. Kondo, MD, Department of Neurology, Juntendo University School of Medicine, 2-1-1 Hongo, Bunkyo-ku, Tokyo 113, Japan.

Subject Index

K. Maurer, P. Riederer,
and H. Beckmann (eds.)

Alzheimer's Disease. Epidemiology, Neuropathology, Neurochemistry, and Clinics

(Key Topics in Brain Research)

1990. 118 figs. (9 in colour). XIX, 581 pages.
Soft cover DM 176,–, öS 1230,–
ISBN 3-211-82197-X

Prices are subject to change without notice

The book "Alzheimer's Disease – Epidemiology, Neuropathology, Neurochemistry, and Clinics" is derived from an International Symposium on the occasion of the 125th Anniversary of Birth of Alois Alzheimer (14. 6. 1864–19. 12. 1915).
Over the past decade, as the elderly have become the fastest growing segment of the population in industrial countries, Alzheimer's disease has emerged as one of major mental health problems. The contributors to this book represent internationally recognized authorities in the field of dementia and present new information about epidemiology, neuropathology, neurochemistry, and clinics in Alzheimer's disease. This book comprises a rich and valuable up-to-date resource for psychiatrists, neurologists, scientists working in the field of neuropathology, neurochemistry and molecular genetics, behavioral scientists, family physicians and all who share an interest in understanding and treating the older individual with Alzheimer's disease/dementia.

Springer-Verlag Wien New York

Horst Przuntek, Peter Riederer (eds.)

Early Diagnosis and Preventive Therapy in Parkinson's Disease

(Key Topics in Brain Research)

1989. 59 figures (1 in color). XIV, 442 pages.
Soft cover DM 135,-, öS 950,-
ISBN 3-211-82080-9

Preisänderungen vorbehalten

At the time when "Parkinson's Disease" is diagnosed in a patient roughly two thirds of dopaminergic neurons of substantia nigra are already degenerated. Therefore, the onset of the disease must be much earlier. This book deals with early diagnosis and early preventive treatment which may sustain the process underlying the disease.

By use of psychometric, kinesiologic, physiological, histological, biochemical, endocrinological, pharmacological and imaging techniques including positronemission tomography and brain mapping specialists tried to focus new diagnostic criteria. New methods including psychometric evaluation, apparative measurement of movement, analysis of peripheral blood and urinary constituents have supplemented this approach. Early preventive therapy has been agreed to consist of low dosis L-DOPA plus benserazide, L-deprenyl and dopaminergic agonists.

Springer-Verlag Wien New York

P. Dostert, P. Riederer, M. Strolin Benedetti, R. Roncucci (eds.)

Early Markers in Parkinson's and Alzheimer's Diseases

(New Vistas in Drug Research, Volume 1)

1990. 38 figures (1 in colour). XX, 310 pages.
Cloth DM 138,–, öS 966,–
ISBN 3-211-82223-2

Prices are subject to change without notice

The aim of "Early Markers in Parkinson's and Alzheimer's Diseases" is to provide the reader with updated data on various approaches whose investigation and development could contribute to the discovery of early diagnostic markers of these two degenerative diseases.

Concerning Parkinson's disease, some of the topics dealt with in the book, will contribute to an updating of the information previously reported in "Early Diagnosis and Preventive Therapy in Parkinson's Disease".

Concerning Alzheimer's disease, the scope and limitations of electrophysiological and brain imaging techniques for an early detection of the disease are documented. Various biochemical parameters, such as brain energy metabolism, levels of choline, and platelet monoamine oxidase activity are envisaged as some of the starting points for the discovery of early diagnostic markers of Alzheimer's disease.

Springer-Verlag Wien New York